CATWATCHING

by the same author

The Biology of Art
Men and Snakes (*co-author*)
Men and Apes (*co-author*)
Men and Pandas (*co-author*)
The Mammals: A Guide to the Living Species
Primate Ethology (*editor*)
The Naked Ape
The Human Zoo
Patterns of Reproductive Behaviour
Intimate Behaviour
Manwatching
Gestures (*co-author*)
Animal Days
The Soccer Tribe
Inrock (*fiction*)
The Book of Ages
The Art of Ancient Cyprus
Bodywatching
The Illustrated Naked Ape
Dogwatching

CATWATCHING

Desmond Morris

JONATHAN CAPE
THIRTY-TWO BEDFORD SQUARE LONDON

Line drawings by Edward Coleridge

First published 1986
Reprinted 1986 (four times), 1987 (three times)

Jonathan Cape Ltd, 32 Bedford Square, London WC1B 3EL

British Library Cataloguing in Publication Data

Morris, Desmond
Catwatching.
1. Cats – Behaviour
I. Title
636.8 SF46.5

ISBN 0-224-02868-5

Typeset by Computape (Pickering) Ltd, North Yorkshire
Printed in Great Britain by
The Alden Press Ltd, Oxford

Contents

Introduction

The domestic cat is a contradiction. No animal has developed such an intimate relationship with mankind, while at the same time demanding and getting such independence of movement and action. The dog may be man's best friend, but it is rarely allowed out on its own to wander from garden to garden or street to street. The obedient dog has to be taken for a walk. The headstrong cat walks alone.

The cat leads a double life. In the home it is an overgrown kitten gazing up at its human owners. Out on the tiles it is fully adult, its own boss, a free-living wild creature, alert and self-sufficient, its human protectors for the moment completely out of mind. This switch from tame pet to wild animal and then back again is fascinating to watch. Any cat-owner who has accidentally come across their pet cat out-of-doors, when it is deeply involved in some feline soap opera of sex and violence, will know what I mean. One instant the animal is totally wrapped up in an intense drama of courtship or status. Then out of the corner of its eye it spots its human owner watching the proceedings. There is a schizoid moment of double involvement, a hesitation, and the animal runs across, rubs against its owner's leg and becomes the house-kitten once more.

The cat manages to remain a tame animal because of the sequence of its upbringing. By living both with other cats (its mother and litter-mates) and with humans (the family that has adopted it) during its infancy and kittenhood, it becomes attached to both and considers that it belongs to both species. It is like a child that grows up in a foreign country and as a consequence becomes bilingual. The cat becomes bi-mental. It may be a cat physically, but mentally it is both feline and human. Once it is fully adult, however, most of its responses are feline ones and it has only one major reaction to its human owners. It treats them as pseudo-parents. This is because they took over from the real mother at a sensitive stage of the kitten's development and went on giving it milk, solid food and comfort as it grew up.

This is rather different from the kind of bond that develops between man and dog. The dog does see its human owners as pseudo-parents, like the cat. On that score the process of attachment is similar. But the dog has an additional link. Canine society is group-organized, feline society is not. Dogs live in packs with tightly controlled status relationships between the individuals. There are top dogs, middle dogs and bottom dogs, and under natural circumstances they move around together, keeping tabs on one another the whole time. So the adult pet dog sees its human family both as pseudo-parents and as the dominant members of its pack. Hence its renowned reputation for obedience and its celebrated capacity for loyalty. Cats do have a complex social organization, but they never hunt in packs. In the wild most of their day is spent in solitary stalking. Going for a walk with a human therefore has no appeal for them. And as for 'coming to heel' and learning to 'sit' and 'stay', they are simply not interested. Such manoeuvres have no meaning for them.

So the moment a cat manages to persuade a human being to open a door (that most hated of human inventions), it is off and away without a backward glance. As it crosses the threshold the cat becomes transformed. The kitten-of-man brain is switched off and the wild-cat brain is clicked on. The dog, in such a situation, may look back to see if its human pack-mate is following to join in the fun of exploring, but not the cat. The cat's mind has floated off into another, totally feline world, where strange bipedal apes have no place.

Because of this difference between domestic cats and domestic dogs, cat-lovers tend to be rather different from dog-lovers. As a rule they have a stronger personality-bias towards independent thought and action. Artists like cats; soldiers like dogs. The much-lauded 'group loyalty' phenomenon is alien to both cats and cat-lovers. If you are a company man, a member of the gang, one of the lads, or picked for the squad, the chances are that at home there is no cat curled up in front of the fire. The ambitious Yuppie, the aspiring politician, the professional footballer, these are not typical cat-owners. It is difficult to picture a rugger-player with a cat in his lap –

much easier to envisage him taking his dog for a walk.

Those who have studied cat-owners and dog-owners as two distinct groups report that there is also a gender-bias. Cat-lovers show a greater tendency to be female. This is not surprising in view of the division of labour that developed during human evolution. Prehistoric males became specialized as group hunters, while the females concentrated on food-gathering and child-rearing. This difference led to a human male 'pack mentality' that is far less marked in females. Wolves, the wild ancestors of domestic dogs, also became pack-hunters, so the modern dog has much more in common with the human male than the human female. An anti-female commentator could refer to women and cats as lacking in team-spirit; an anti-male one to men and dogs as gangsters.

The argument will always go on – feline self-sufficiency and individualism versus canine camaraderie and good-fellowship. But it is important to stress that in making a valid point I have caricatured the two positions. In reality there are many people who enjoy equally the company of both cats and dogs. And all of us, or nearly all of us, have both feline and canine elements in our personalities. We have moods when we want to be alone and thoughtful, and other times when we wish to be in the centre of a crowded, noisy room.

Both the cat and the dog are animals with which we humans have entered into a solemn contract. We made an unwritten, unspoken pact with their wild ancestors, offering food and drink and protection in exchange for the performance of certain duties. For dogs, the duties were complex, involving a whole range of hunting tasks, as well as guarding property, defending their owners against attack, destroying vermin, and acting as beasts of burden pulling our carts and sledges. In modern times an even greater range of duties has been given to the patient, long-suffering canine, including such diverse activities as guiding the blind, trapping criminals and running races.

For cats, the terms of the ancient contract were much simpler and have always remained so. There was just one primary task and one secondary one. They were required to act firstly as pest-controllers and then, in addition, as household pets. Because they are solitary hunters of small prey they were of little use to human huntsmen in the field. Because they do not live in tightly organized social groups depending on mutual aid to survive, they do not raise the alarm in response to intruders in the home, so they were little use as guardians of property, or as defenders of their owners. Because of their small size they could offer no assistance whatever as beasts of burden. In modern times, apart from sharing the honours with dogs as the ideal house-pets, and occasionally sharing the acting honours in films and plays, cats have failed to diversify their usefulness to mankind.

Despite this narrower involvement in human affairs, the cat has managed to retain its grip on human affections. There are almost as many cats as dogs in the British Isles, according to recent estimates: about five million cats to six million dogs. In the United States the ratio is slightly less favourable to felines: about twenty-three million cats to forty million dogs. Even so, this is a huge population of domestic cats and, if anything, it is probably an underestimate. Although there are still a few mousers and ratters about, performing their ancient duties as vermin destroyers, the vast majority of all domestic cats today are family pets or feral survivors. Of the family pets, some are pampered pedigrees but most are moggies of mixed ancestry. The proportion of pedigree cats to moggies is probably lower than that for pedigree dogs to mongrels. Although cat shows are just as fiercely contested as dog shows, there are fewer of them, just as there are fewer breeds of show cats. Without the wide spectrum of ancient functions to fulfil, there was far less breed specialization in the early days. Indeed, there was hardly any. All breeds of cat are good mousers and ratters, and no more was required of them. So any modifications in coat length, colour or pattern, or in body proportions, had to arise purely on the basis of local preferences and owners' whims. This has led to some strikingly beautiful pedigree cat breeds, but not to the amazing range of dramatically different types found among dogs. There is no cat equivalent of the Great Dane or the Chihuahua, the St Bernard or the Dachshund. There is a good deal of variation in fur type and colour, but very little in body shape and size. A really large cat weighs in at around eighteen pounds; the smallest at about three pounds. This means that, even when considering almost freakish feline extremes, big domestic cats are only six times as heavy as small ones, compared with the situation among dogs, where a St Bernard can weigh 300 times as much as a little Yorkshire Terrier. In other words, the weight variation of dogs is fifty times as great as in cats.

Turning to abandoned cats and those that have gone wild through choice – the feral population – one also notices a considerable difference. Stray dogs form self-supporting packs and start to breed and fend for themselves without human aid in less civilized regions, but such groups have become almost non-existent in urban and suburban areas. Indeed, in modern, crowded European countries they hardly exist anywhere. Even the rural districts cannot support them. If a feral pack forms, it is soon hunted down by the farming community to prevent attacks on their stock. Feral cat colonies are another matter. Every major city has a thriving population of them. Attempts to eradicate them usually fail because there are always new strays to add to the pool. And the need to destroy them is not felt to be so great, because they can often survive by continuing their age-old function of pest control. Where human intervention has eliminated the rat and mouse population by poisoning, however, the feral cats must live on their wits, scavenging from dustbins and begging from soft-hearted humans. Many of these back-alley cats are pathetic creatures on the borderline of survival. Their resilience is amazing and a testimony to the fact that, despite millennia of domestication, the feline brain and body are still remarkably close to the wild condition.

At the same time this resilience is to blame for a great deal of feline suffering. Because cats *can* survive when thrown out and abandoned, it makes it easier for people to do just that. The fact that most of these animals must then live out their years in appalling urban conditions – slum cats scratching a living among the garbage and filth of human society – may reflect how tough they are, but it is a travesty of feline existence. That we tolerate it is one more example of the shameful manner in which we have repeatedly broken our ancient contract with the cat. It is nothing, however, compared with the brutal way we have sometimes tormented and tortured cats over the centuries. They have all too frequently been the butt of our redirected aggression, so much so that we even have a popular saying to express the phenomenon: '. . . and the office boy kicked the cat', illustrating the way in which insults from above become diverted to victims lower down the social order, with the cat at the bottom of the ladder.

Fortunately, against this can be set the fact that the vast majority of human families owning pet cats do treat their animals with care and respect. The cats have a way of endearing themselves to their owners, not just by their 'kittenoid' behaviour, which stimulates strong parental feelings, but also by their sheer gracefulness. There is an elegance and a composure about them that captivates the human eye. To the sensitive human being it becomes a privilege to share a room with a cat, exchange its glance, feel its greeting rub, or watch it gently luxuriate itself into a snoozing ball on a soft cushion. And for millions of lonely people – many physically incapable of taking long walks with a demanding dog – the cat is the perfect companion. In particular, for people forced to live on their own in later life, their company provides immeasurable rewards. Those tight-lipped puritans who, through callous indifference and a sterile selfishness, seek to stamp out all forms of pet-keeping in modern society would do well to pause and consider the damage their actions may cause.

This brings me to the purpose of *Catwatching*. As a zoologist, I have had in my care, at one time or another, most members of the cat family, from great Tigers to tiny Tiger-Cats, from powerful Leopards to diminutive Leopard-Cats, and from mighty Jaguars to rare little Jaguarondis. At home there has nearly always been a domestic moggie to greet my return, sometimes with a cupboard-full of kittens. As a boy growing up in the Wiltshire countryside, I spent many hours lying in the grass, observing the farm cats as they expertly stalked their prey, or spying on the hayloft nests where they suckled their squirming kittens. I developed the habit of catwatching early in life, and it has stayed with me now for nearly half a century. Because of my professional involvement with animals I have frequently been asked questions about cat behaviour, and I have been surprised at how little most people seem to know about these intriguing animals. Even those who dote on their own pet cat often have only a vague understanding of the complexities of its social life, its sexual behaviour, its aggression or its hunting skills. They know its moods well and care for it fastidiously, but they do not go out of their way to study their pet. To some extent this is not their fault, as much feline behaviour takes place away from the home base of the kitchen and the living room. So I hope that even those who feel they know their own cats intimately may learn a little more about their graceful companions by reading these pages.

The method I have used is to set out a series of basic questions and then to provide simple, straightforward answers to them. There are plenty of good, routine books on cat care, which give all the usual details about feeding, housing and veterinary treatment, combined with a classificatory list of the various cat breeds and their characteristics. I have not repeated those details here. Instead I have tried to provide a different sort of cat book, one that concentrates on feline behaviour and gives replies to the sort of queries with which I have been confronted over the years. If I have succeeded, then, the next time you encounter a cat, you should be able to view the world in a more feline way. And once you have started to do that, you will find yourself asking more and more questions about their fascinating world, and perhaps you too will develop the urge to do some serious catwatching.

The Cat

We know for certain that 3,500 years ago the cat was already fully domesticated. We have records from ancient Egypt to prove this. But we do not know when the process began. The remains of cats have been found at a neolithic site at Jericho dating from 9,000 years ago, but there is no proof that those felines were domesticated ones. The difficulty arises from the fact that the cat's skeleton changed very little during its shift from wild to tame. Only when we have specific records and detailed pictures – as we do from ancient Egypt – can we be sure that the transformation from wild cat to domestic animal had taken place.

One thing is clear: there would have been no taming of the cat before the Agricultural Revolution (in the neolithic period, or New Stone Age). In this respect the cat differed from the dog. Dogs had a significant role to play even before the advent of farming. Back in the palaeolithic period (or Old Stone Age), prehistoric human hunters were able to make good use of a four-legged hunting companion with superior scenting abilities and hearing. But cats were of little value to early man until he had progressed to the agricultural phase and was starting to store large quantities of food. The grain stores, in particular, must have attracted a teeming population of rats and mice almost from the moment that the human hunter settled down to become a farmer. In the early cities, where the stores were great, it would have become an impossible task for human guards to ambush the mice and kill them in sufficient numbers to stamp them out or even to prevent them from multiplying. A massive infestation of rodents must have been one of the earliest plagues known to urban man. Any carnivore that preyed on these rats and mice would have been a godsend to the harassed food-storers.

It is easy to visualize how one day somebody made the casual observation that a few wild cats had been noticed hanging around the grain stores, picking off the mice. Why not encourage them? For the cats, the scene must have been hard to believe. There all around them was a scurrying feast on a scale they had never encountered before. Gone were the interminable waits in the undergrowth. All that was needed now was a leisurely stroll in the vicinity of the vast grain stores and a gourmet supermarket of plump, grain-fed rodents awaited them. From this stage to the keeping and breeding of cats for increased vermin destruction must have been a simple step, since it benefited both sides.

With our efficient modern pest-control methods it is difficult for us to imagine the significance of the cat to those early civilizations, but a few facts about the attitudes of the ancient Egyptians towards their beloved felines will help to underline the importance that was placed upon them. They were, for instance, considered sacred, and the punishment for killing one was death. If a cat happened to die naturally in a house, all the human occupants had to enter full mourning, which included the shaving-off of their eyebrows.

Following death, the body of an Egyptian cat was embalmed with full ceremony, the corpse being bound in wrappings of different colours and its face covered with a sculptured wooden mask. Some were placed in cat-shaped wooden coffins, others were encased in plaited straws. They were buried in enormous feline cemeteries in huge numbers – literally millions of them.

The cat-goddess was called Bastet, meaning She-of-Bast. Bast was the city where the main cat temple was situated, and where each spring as many as half a million people converged for the sacred festival. About 100,000 mummified cats were buried at each of these festivals to honour the feline virgin-goddess (who was presumably a forerunner of the Virgin Mary). These Bastet festivals were said to be the most popular and best attended in the whole of ancient Egypt, a success perhaps not unconnected with the fact that they included wild orgiastic celebrations and 'ritual frenzies'. Indeed, the cult of the cat was so popular that it lasted for nearly 2,000 years. It was officially banned in AD 390, but by then it was already in serious decline. In its heyday, however, it reflected the immense esteem in which the cat was held in that ancient civilization, and the many beautiful bronze statues of cats that have survived bear testimony to the Egyptians' appreciation of its graceful form.

A sad contrast to the ancient worship of the cat is the vandalizing of the cat cemeteries by the British in the last century. One example will suffice: a consignment of 300,000 mummified cats was shipped to Liverpool where they were ground up for use as fertilizer on the fields of local farmers. All that survives from this episode is a single cat skull which is now in the British Museum.

The early Egyptians would probably have demanded 300,000 deaths for such sacrilege, having once torn a Roman soldier limb from limb for hurting a cat. They not only worshipped their cats, but also expressly prohibited their export. This led to repeated attempts to smuggle them out of the country as high-status house pets. The Phoenicians, who were the ancient equivalent of second-hand car salesmen, saw catnapping as an intriguing challenge and were soon shipping out high-priced moggies to the jaded rich all around the Mediterranean. This may have annoyed the Egyptians, but it was good news for the cat in those early days, because it introduced them to new areas as precious objects to be well treated.

Plagues of rodents that were sweeping Europe gave the cat a new boost as a pest-controller, and it rapidly spread across the continent. The Romans were largely responsible for this, and it was they who brought the cat to Britain. We know that cats were well treated in the centuries that followed because of the punishments that are recorded for killing one. These were not as extreme as in ancient Egypt, but fines of a lamb or a sheep were far from trivial. The penalty devised by one Welsh king in the tenth century reflected the significance to him of the dead cat. The animal's corpse was suspended by its tail with its nose touching the ground, and the punishment for its killer was to heap grain over the body until it disappeared beneath the mound. The confiscation of this grain gave a clear picture of how much a working cat was estimated to save from the bellies of rats and mice.

These good times for cats were not to last, however. In the Middle Ages the feline population of Europe was to experience several centuries of torture, torment and death at the instigation of the Christian Church. Because they had been involved in earlier pagan rituals, cats were proclaimed evil creatures, the agents of Satan and familiars of witches. Christians everywhere were urged to inflict as much pain and suffering on them as possible. The sacred had become the damned. Cats were publicly burned alive on Christian feast days. Hundreds of thousands of them were flayed, crucified, beaten, roasted and thrown from the tops of church towers at the urging of the priesthood, as part of a vicious purge against the supposed enemies of Christ.

Happily, the only legacy we have today of that miserable period in the history of the domestic cat is the surviving superstition that a black cat is connected with luck. The connection is not always clear, however, because as you travel from country to country the luck changes from good to bad, causing much confusion. In Britain, for instance, a black cat means good luck, whereas in America and continental Europe it usually means bad luck. In some regions this superstitious attitude is still taken remarkably seriously. For example, a few years ago a wealthy restaurant-owner was driving to his home south of Naples late one night when a black cat ran across the road in front of his car. He stopped and pulled in to the side of the road, unable to proceed unless the cat returned (to 'undo' the bad luck). Seeing him parked there on the lonely road late at night, a cruising police car pulled up and the officers questioned him. When they learned the reason, such was the strength of cat superstition that, refusing to drive on for fear of bringing bad luck on themselves, they also had to sit in their car and wait for the cat to return.

Although these superstitions still survive, the cat is now once again the much-loved house pet that it was in ancient Egypt. It may not be sacred, but it is greatly revered. The Church's cruel persecution has long since been rejected by ordinary people and, during the nineteenth century, a new phase of cat promotion exploded in the shape of competitive cat shows and pedigree cat breeding.

As already mentioned, the cat had not been bred into many different forms for different work tasks, like the dog, but there had been a number of local developments, with variations in colour, pattern and coat length arising almost accidentally in different countries. Travellers in the nineteenth century started to collect the strange-looking cats they met abroad and to transport them back to Victorian England. There they bred them carefully to intensify their special characteristics. Cat shows became increasingly popular, and during the past 150 years more than 100 different pedigree breeds have been standardized and registered in Europe and North America.

All these modern breeds appear to belong to one species: *Felis sylvestris*, the Wild Cat, and are capable of interbreeding, both with one another and with all races of the wild *sylvestris*. At the very start of feline domestication, the Egyptians began by taming the North African race of *Felis sylvestris*. Until recently this was thought of as a distinct species and was called *Felis lybica*. It is now known to be no more than a race and is designated as *Felis sylvestris lybica*. It is smaller and more slender than the European race of Wild Cat and was apparently easier to tame. But when the Romans progressed through Europe, bringing their domestic cats with them, some of the animals mated with the stockier northern Wild Cats and produced heavier, more robust offspring. Today's modern cats reflect this, some being big and sturdy, like many of the tabbies, while others are more elongated and angular, like the various Siamese breeds. It is likely that these Siamese animals and the other more slender breeds are closer to the Egyptian original, their domestic ancestors having been dispersed throughout the world without any contact with the heavy-set northern Wild Cats.

Although opinions have differed, it now seems highly improbable that any other species of wild feline was involved in the history of modern domestic cats. We do know that a second, bigger cat, *Felis chaus*, the Jungle Cat, was popular with the ancient Egyptians, but it appears to have dropped out of the running very early on. We can, however, be certain that it was originally a serious contender for the domestication stakes, because examination of mummified cats has revealed that some of them possessed the much larger Jungle Cat skulls. But although the Jungle Cat is one of the more friendly cats in captivity, it is huge in comparison with even the heftiest of modern domestic animals and it is therefore unlikely that it played a part in the later domestication story.

This is not the place to give details of the modern cat breeds, but a brief history of their introduction will help to give some idea of the way the modern cat 'fancy' has become established:

The oldest breeds involved are the various Shorthaired Cats, descendants of the animals spread across Europe by the Romans. There is then a long gap until the sixteenth century, when ships from the Orient arrived at the Isle of Man carrying a strange tail-less cat – the famous Manx. Because of its curiously mutilated appearance, this breed has never been widely popular, though it still has its devotees. At about the same time the first of the Longhaired Cats, the beautiful Angora, was brought into Europe from its Turkish homelands. Later, in the mid-nineteenth century, it was to be largely eclipsed by the even more spectacular Persian Cat from Asia Minor, with its enormously thick, luxuriant coat of fur.

Then, in the late nineteenth century, in complete contrast, the angular, elongated Siamese arrived from the Far East. With its unique personality – far more extrovert than other cats – it appealed to a quite different type of cat-owner. Whereas the Persian was the perfect, rounded, fluffy child substitute with a rather infantile, flattened face, the Siamese was a much more active companion.

At about the same time as the appearance of the Siamese, the elegant Russian Blue was imported from Russia and the tawny, wild-looking Abyssinian from what is now Ethiopia.

In the present century, the dusky Burmese was taken to the United States in the 1930s and from there to Europe. In the 1960s several unusual additions appeared as sudden mutations: the bizarre Sphynx, a naked cat from Canada; the crinkly-haired Rex from Devon and Cornwall; and the flattened-eared Fold Cat from Scotland. In the 1970s the Japanese Bobtail Cat, with its curious little stump making it look like a semi-Manx Cat, was imported into the United States; the crinkly Wire-haired Cat was developed from a mutation in America; and the diminutive Drain-Cat (so called because drains were a good place to hide in cat-scorning Singapore) appeared on the American scene, rejoicing in the exotic name of Singapura.

Finally there was the extraordinary Ragdoll Cat, with the strangest temperament of any feline. If picked up, it hangs limply like a rag doll. It is so placid that it gives the impression of being permanently drugged. Nothing seems to worry it. More of a hippie-cat than a hip-cat, it seems only appropriate that it was first bred in California.

This is by no means an exhaustive list, but it gives some idea of the range of cats available to the pedigree enthusiast. With many of the breeds I have mentioned there is a whole range of varieties and colour-types, dramatically increasing the list of show categories. Each time a new type of cat appears the fur flies – not from fighting felines, but from the unseemly skirmishes that break out between the overenthusiastic breeders of the new line and the unduly autocratic authorities that govern the major cat shows. Latest breed to top the tussle-charts is the aforementioned Ragdoll: ideal for invalids, say its defenders; too easy to injure, say its detractors.

To add to the complications, there are considerable disagreements between the different show authorities, with the Governing Council of the Cat Fancy in Britain recognizing different breeds from the Cat Fanciers' Association in America, and the two organizations sometimes confusingly giving different names to the same breed. None of this does much harm, however. It simply has the effect of adding the excitement of a great deal of heated argument and debate, while the pedigree cats themselves benefit from all the interest that is taken in them.

The seriousness with which competitive cat-showing is treated also helps to raise the status of all cats, so that the ordinary pet moggies benefit too in the long run. And they remain the vast majority of all modern domestic cats because, to most people, as Gertrude Stein might have said, a cat is a cat is a cat. The differences, fascinating though they are, remain very superficial. Every single cat carries with it an ancient inheritance of amazing sensory capacities, wonderful vocal utterances and body language, skilful hunting actions, elaborate territorial and status displays, strangely complex sexual behaviour and devoted parental care. It is an animal full of surprises, as we shall see on the pages that follow.

Why does a cat purr?

The answer seems obvious enough. A purring cat is a contented cat. This surely must be true. But it is not. Repeated observation has revealed that cats in great pain, injured, in labour and even dying often purr loud and long. These can hardly be called contented cats. It is true, of course, that contented cats do also purr, but contentment is by no means the sole condition for purring. A more precise explanation, which fits all cases, is that purring signals a friendly social mood, and it can be given as a signal to, say, a vet from an injured cat indicating the *need* for friendship, or as a signal to an owner, saying thank you for friendship given.

Purring first occurs when kittens are only a week old and its primary use is when they are being suckled by their mother. It acts then as a signal to her that all is well and that the milk supply is successfully reaching its destination. She can lie there, listening to the grateful purrs, and know without looking up that nothing has gone amiss. She in turn purrs to her kittens as they feed, telling them that she too is in a relaxed, co-operative mood. The use of purring among adult cats (and between adult cats and humans) is almost certainly secondary and is derived from this primal parent-offspring context.

An important distinction between small cats, like our domestic species, and the big cats, like lions and tigers, is that the latter cannot purr properly. The tiger will greet you with a friendly 'one-way purr' – a sort of juddering splutter – but it cannot produce the two-way purr of the domestic cat, which makes its whirring noise not only with each outward breath (like the tiger), but also with each inward breath. The exhalation/inhalation rhythm of feline purring can be performed with the mouth firmly shut (or full of nipple), and may be continued for hours on end if the conditions are right. In this respect small cats are one up on their giant relatives, but big cats have another feature which compensates for it – they can roar, which is something small cats can never do.

Why does a cat like being stroked?

Because it looks upon humans as 'mother cats'. Kittens are repeatedly licked by their mothers during their earliest days and the action of human stroking has much the same feel on the fur as feline licking. To the kitten, the mother cat is 'the one who feeds, cleans and protects'. Because humans continue to do this for their pets long after their kitten days are behind them, the domesticated animals never fully grow up. They may become full-sized and sexually mature, but in their minds they remain kittens in relation to their human owners.

For this reason cats – even elderly cats – keep on begging for maternal attention from their owners, pushing up to them and gazing at them longingly, waiting for the pseudo-maternal hand to start acting like a giant tongue again, smoothing and tugging at their fur. One very characteristic body action they perform when they are being stroked, as they greet their 'mothers', is the stiff erection of their tails. This is typical of young kittens receiving attention from their real mothers and it is an invitation to her to examine their anal regions.

Why does a cat tear at the fabric of your favourite chair?

The usual answer is that the animal is sharpening its claws. This is true, but not in the way most people imagine. They envisage a sharpening-up of blunted points rather in the manner of humans improving the condition of blunted knives. But what really occurs is the stripping-off of the old, worn-out claw sheaths to reveal glistening new claws beneath. It is more like the shedding of a snake's skin than the sharpening of a kitchen knife. Sometimes, when people run their hands over the place where the cat has been tearing at the furniture, they find what they think is a ripped-out claw and they then fear that their animal has accidentally caught its claw in some stubborn threads of the fabric and damaged its foot. But the 'ripped-out claw' is nothing more than the old outer layer that was ready to be discarded.

Cats do not employ these powerful 'stropping' actions with the hind feet. Instead they use their teeth to chew off the old outer casings from the hind claws.

A second important function of the stropping with the front feet is the exercising and strengthening of the retraction and protrusion apparatus of the claws, so vital in catching prey, fighting rivals and climbing.

A third function, not suspected by most people, is that of scent-marking. There are scent glands on the underside of the cat's front paws and these are rubbed vigorously against the fabric of the furniture being clawed. The rhythmic stropping, left paw, right paw, squeezes scent on to the surface of the cloth and rubs it in, depositing the cat's personal signature on the furniture. This is why it is always your favourite chair which seems to suffer most attention, because the cat is responding to your own personal fragrance and adding to it. Some people buy an expensive scratching-post from a pet shop, carefully impregnated with catnip to make it appealing, and are bitterly disappointed when the cat quickly ignores it and returns to stropping the furniture. Hanging an old sweat-shirt over the scratching-post might help to solve the problem, but if a cat has already established a particular chair or a special part of the house as its 'stropping spot', it is extremely hard to alter the habit.

In desperation, some cat-owners resort to the cruel practice of having their pets de-clawed. Apart from the physical pain this inflicts, it is also psychologically damaging to the cat and puts it at a serious disadvantage in all climbing pursuits, hunting actitivies and feline social relationships. A cat without its claws is not a true cat.

Why does a cat roll over to lie on its back when it sees you?

When you enter a room where a cat is lying asleep on the floor and you greet it with a few friendly words, it may respond by rolling over on its back, stretching out its legs as far as they will go, yawning, exercising its claws and gently twitching the tip of its tail. As it performs these actions, it stares at you, checking your mood. This is a cat's way of offering you a passively friendly reaction and it is something which is only done to close family intimates. Few cats would risk such a greeting if the person entering the room were a stranger, because the belly-up posture makes the animal highly vulnerable. Indeed, this is the essence of its friendliness. The cat is saying, in effect, 'I roll over to show you my belly to demonstrate that I trust you enough to adopt this highly vulnerable posture in your presence.'

A more active cat would rush over to you and start rubbing against you as a form of friendly greeting, but a cat in a lazy, sleepy mood prefers the belly-roll display. The yawning and stretching that accompany it reflect the sleepiness of the animal – a sleepiness which it is prepared to interrupt just so much and no more. The slight twitching of the tail indicates that there is a tiny element of conflict developing – a conflict between remaining stretched out and jumping up to approach the new arrival.

It is not always safe to assume that a cat making this belly-up display is prepared to allow you to stroke its soft underside. It may appear to be offering this option, but frequently an attempt to respond with a friendly hand is met with a swipe from an irritated paw. The belly region is so well protected by the cat that it finds contact there unpleasant, except in relationships where the cat and its human owner have developed a very high degree of social intimacy. Such a cat may trust its human family to do almost anything to it. But the more typical, wary cat draws the line at approaches to its softer parts.

Why does a cat rub up against your leg when it greets you?

Partly to make friendly physical contact with you, but there is more to it than that. The cat usually starts by pressing against you with the top of its head or the side of its face, then rubs all along its flank and finally may slightly twine its tail around you. After this it looks up and then repeats the process, sometimes several times. If you reach down and stroke the animal, it increases its rubbing, often pushing the side of its mouth against your hand, or nudging upwards with the top of its head. Then eventually it wanders off, its greeting ritual complete, sits down and washes its flank fur.

All these elements have special meanings. Essentially what the cat is doing is implementing a scent-exchange between you and it. There are special scent glands on the temples and at the gape of the mouth. Another is situated at the root of the tail. Without your realizing it, your cat has marked you with its scent from these glands. The feline fragrances are too delicate for our crude noses, but it is important that friendly members of the cat's family should be scent-sharing in this way. This makes the cat feel more at home with its human companions. And it is important, too, for the cat to read *our* scent signals. This is achieved by the flank-rubbing element of the greeting, followed by the cat sitting down and 'tasting' us with its tongue – through the simple process of licking the fur it has just rubbed so carefully against us.

Why do some cats hop up on their hind legs when greeting you?

One of the problems cats have when adjusting to human companions is that we are much too tall for them. They hear our voices coming from what is, to them, a great height and they find it hard to greet such a giant in the usual way. How can they perform the typical cat-to-cat greeting of rubbing faces with one another? The answer is that they cannot. They have to make do with rubbing our legs or a downstretched hand. But it is in their nature to aim their greetings more towards the head region, and so they make a little intention movement of doing this – the stiff-legged hop in which the front feet are lifted up off the ground together, raising the body for a brief moment before letting it fall back again to its usual four-footed posture. This greeting hop is therefore a token survival of a head-to-head contact.

A clue to this interpretation comes from the way small kittens sometimes greet their mother when she returns to the nest. If they have developed to the point where their legs are strong enough for the 'hop', the kittens will peform a modest version of the same movement, as they push their heads up towards that of the mother cat. In their case there is not far to go, and she helps by lowering her own head towards theirs, but the incipient hop is clear enough.

As with all rubbing-greetings, the head-to-head contact is a feline method of mingling personal scents and turning them into shared family scents. Some cats use their initiative to re-create a better head contact when greeting their human friends. Instead of the rather sad little symbolic hop, they leap up on to a piece of furniture near the human and employ this elevated position to get themselves closer for a more effective face-to-face rub.

Why does a cat trample on your lap with its front paws?

All cat-owners have experienced the moment when their cat jumps up and with cautious movements settles itself down on their lap. After a short pause it starts to press down, first with one front paw and then with the other, alternating them in a rhythmic kneading or trampling action. The rhythm is slow and deliberate as if the animal is marking time in slow motion. As the action becomes more intense the prick of claws can be felt, and at this point the owner usually becomes irritated and shoos the cat away, or gently picks it up and places it on the floor. The cat is clearly upset by this rebuff and the owners are similarly put out when, brushing away a few cat hairs, they discover that the animal has been dribbling while trampling. What does all this mean?

To find the answer it is necessary to watch kittens feeding at the nipple. There the same actions can be observed, with the kittens' tiny paws kneading away at their mother's belly. These are the movements which stimulate the flow of milk to the nipples and the dribbling is part of the mouthwatering anticipation of delicious nourishment to come. This 'milk-treading', as it is called, is done at a very slow pace of approximately one stroke every two seconds, and it is always accompanied by loud purring. What happens when the adult pet tramples on the lap of its human owner must therefore be interpreted as a piece of infantile behaviour. It would seem that when the owner sits down in a relaxed manner, signals are given off saying to the cat, 'I am your mother lying down ready to feed you at the breast.' The adult cat then proceeds to revert to kittenhood and squats there, purring contentedly and going through the motions of stimulating a milk supply.

From the cat's point of view this is a warm, loving moment and its bodily removal by a claw-pricked owner must be quite inexplicable. No good cat mother would behave in such a negative way. People react rather differently. To the cat they are clearly maternal figures, because they do supply milk (in a saucer) and other nourishment, and they do sit down showing their undersides in an inviting manner, but once the juvenile reaction of milk-treading is given, they suddenly and mystifyingly become upset and thrust the pseudo-infant from them.

This is a classic example of the way in which interactions between humans and cats can lead to misunderstandings. Many can be avoided by recognizing the fact that an adult domestic cat remains a kitten in its behaviour towards its pseudo-parental owner.

Why does a cat bury its faeces?

This action is always referred to as an indication of the fastidious tidiness of the cat. Owners of messy dogs are often regaled with this fact by cat-owners insisting on the superiority of felines over canines. This favoured interpretation of faeces-burying as a sign of cat hygiene does not, however, stand up to close investigation.

The truth is that cats bury their faeces as a way of damping down their odour display. Faeces-burying is the act of a subordinate cat, fearful of its social standing. Proof of this was found when the social lives of feral cats were examined closely. It was discovered that dominant tom-cats, far from burying their faeces, actually placed them on little 'advertising' hillocks, or any other raised points in the environment where the odour could be wafted abroad to maximum effect. It was only the weaker, more subdued cats which hid their faeces. The fact that our pet cats always seem to carry out such a careful burying routine is a measure of the extent to which they see themselves dominated by us (and also perhaps by the other cats in the neighbourhood). This is not really so surprising. We are physically stronger than they are and we completely dominate that all-important element in feline life – the food supply. Our dominance is in existence from the time of kittenhood onwards, and never in serious doubt. Even big cats, such as lions, can be kept in this subordinate role for a lifetime, by their friendly owners, so it is hardly surprising that the small domestic cat is permanently in awe of us and therefore always makes sure to bury its faeces.

Burying the faeces does not, of course, completely switch off the odour signal, but it does reduce it drastically. In this way the cat can continue to announce its presence through its scents, but not to the extent that it transmits a serious threat.

Why does a cat spend so much time grooming its fur?

The obvious answer is to keep itself clean, but there is much more to grooming than this. In addition to cleaning away dust and dirt or the remains of the last meal, the repeated licking of the fur helps to smooth it so that it acts as a more efficient insulating layer. A ruffled coat is a poor insulator, which can be a serious hazard for a cat in freezing weather.

Cold is not the only problem. Cats easily overheat in summer-time and fur-grooming increases then for a special reason. Cats do not have sweat glands all over their bodies as we do, so they cannot use sweating as a rapid method of cooling. Panting helps, but it is not enough. The solution is to lick repeatedly at the fur and deposit on it as much saliva as possible. The evaporation of this saliva then acts in the same way as the evaporation of sweat on our skin.

If cats have been in sunlight they increase their grooming even more. This is not, as might be imagined, simply because they are even hotter, but because the action of sunlight on their fur produces essential vitamin D. They acquire this crucial additive to their diet by the licking movements of their tongues over the sun-warmed fur.

Grooming also increases when cats become agitated. This is called displacement grooming and it is believed to act as an aid to relieving the strain of tense social encounters. When we are in a state of conflict we often 'scratch our heads'. A cat under similar conditions licks its fur.

Any cat-owner who has just been holding or cuddling their cat will be familiar with the animal's actions as soon as it has been released from human contact. It wanders off, sits down and then, nearly always, starts to groom itself. This is partly because it needs to smooth its ruffled fur, but there is also another reason. You have, by handling the cat, given it your scent and to some extent masked the cat's scent. The licking of the fur redresses the balance, weakening your scent and reinforcing the cat's own odour on its body surface. Our lives are dominated by visual signals, but in the cat's world odours and fragrances are much more important, and an overdose of human scent on its fur is disturbing and has to be rapidly corrected. In addition, the licking of the fur you have been handling means that the cat can actually enjoy 'tasting' you and reading the signals it gets from the scent of your sweat glands. *We* may not be able to smell the odour of our hands, but a cat can.

Finally, the vigorous tugging at the fur which is so typical of a cat's self-grooming actions plays a special role in stimulating the skin glands at the base of the individual hairs. The secretions of these glands are vital to keep the fur waterproofed, and the tugging of the cat's busy tongue steps up the waterproofing as a protection against the rain.

So grooming is much more than mere cleaning. When it licks its fur a cat is protecting itself, not only from dirt and disease, but also from exposure to cold, from overheating, from vitamin deficiency, from social tension, from foreign odours and from getting drenched to the skin. No wonder it devotes so much of its waking day to this piece of behaviour.

There is one danger inherent in this activity. Moulting cats and cats with very long fur quickly accumulate a large number of hairs inside their alimentary tracts and these form into hairballs which can cause obstructions. Usually hairballs are vomited up naturally without causing any trouble, but if they grow too large they may become a serious hazard. Cats of a nervous disposition, which do a great deal of displacement grooming, also suffer in this way. To solve their problem it is necessary to find out what is causing their agitation and deal with it. For the moulting and long-haired cats the only prevention is regular grooming by the cat's owner with brush and comb, to remove the excess fur.

Self-grooming begins when the kitten is about three weeks old, but it has its fur attended to by the mother from the moment of its birth. Being groomed by another cat is called allogrooming, in contrast with self-grooming which is known technically as autogrooming. Allogrooming is common not only between mother and kitten, but also between adult cats that have grown up together and have developed a close social bond. Its primary function is not mutual hygiene, but rather a cementing of the friendly relation that exists between the two animals. All the same, licking in a region that is hard for the cat itself to reach does have a special appeal, and cats are partial to attention behind the ears. This is why tickling and rubbing behind the ears is such a popular form of contact between cat-owners and their cats.

The autogrooming actions often follow a set sequence, when a cat is indulging in a complete 'wash-and-brush-up'. The typical routine goes as follows:

1. Lick the lips.
2. Lick the side of one paw until it is wet.
3. Rub the wet paw over the head, including ear, eye, cheek and chin.
4. Wet the other paw in the same way.
5. Rub the wet paw over that side of the head.
6. Lick front legs and shoulders.
7. Lick flanks.
8. Lick genitals.
9. Lick hind legs.
10. Lick tail from base to tip.

If at any stage during this process an obstruction is encountered – a tangled bit of fur, for example – the licking is momentarily abandoned in favour of a localized nibble with the teeth. Then, when all is clear, the grooming sequence is resumed. Foot and claw nibbling are particularly common, removing dirt and improving the condition of the claws. This complicated cleaning sequence differs from that seen in many other mammals. Rats and mice, for example, use the whole of their front paws for grooming their heads, whereas the cat uses only the side of the paw and part of the forearm. Also, rodents sit up on their back legs and groom with both front feet at the same time, while the feline technique is to employ each front leg alternately, resting its body on the one not in use. Human observers rarely comment on such differences, remarking simply that an animal is busy cleaning itself. In reality, closer observation reveals that each species follows a characteristic and complex sequence of actions.

Why does a cat wag its tail?

Most people imagine that if a cat wags its tail it must be angry, but this is only a partial truth. The real answer is that the cat is in a state of conflict. It wants to do two things at once, but each impulse blocks the other. For example, if a cat cries to be let out at night and the door is opened to reveal a downpour of heavy rain, the animal's tail may start to wag. If it rushes out into the night and stands there defiantly for a moment, getting drenched, its tail wags even more furiously. Then it makes a decision and either rushes back in to the comforting shelter of the house, or bravely sets off to patrol its territory, despite the weather conditions. As soon as it has resolved its conflict, one way or the other, its tail immediately stops wagging.

In such a case it is inappropriate to describe the mood as one of anger. Anger implies a frustrated urge to attack, but the cat in the rainstorm is not aggressive. What is being frustrated there is the urge to explore, which in turn is frustrating the powerful feline desire to keep snug and dry. When the two urges momentarily balance one another, the cat can obey neither. Pulled in two different directions at once, it stands still and wags its tail. Any two opposing urges would produce the same reaction, and only when one of these was the urge to attack – frustrated by fear or some other competing mood – could we say that the cat was wagging its tail because it was angry.

If tail-wagging in cats represents a state of acute conflict, how did such a movement originate? To understand this, watch a cat trying to balance on a narrow ledge. If it feels itself tipping over, its tail quickly swings sideways, acting as a balancing organ. If you hold a cat on your lap and then tip it slightly to the left and then to the right, alternating these movements, you can see its tail swinging rhythmically from side to side as if it is wagging it in slow motion. This is how the tail-wagging movement used in mood-conflicts began. As the two competing urges pulled the cat in opposite directions, the tail responded as if the animal's body were being tipped over first one way and then the other. During evolution this lashing of the tail from side to side became a useful signal in the body language of cats and was greatly speeded up in a way that made it more conspicuous and instantly recognizable. Today it is so much faster and more rhythmic than any ordinary balancing movement of the tail that it is easy to tell at a glance that the conflict the animal is experiencing is emotional rather than purely physical.

Why does a tom-cat spray urine on the garden wall?

Tom-cats mark their territories by squirting a powerful jet of urine backwards on to vertical features of their environment. They aim at walls, bushes, tree-stumps, fence-posts or any landmark of a permanent kind. They are particularly attracted to places where they or other cats have sprayed in the past, adding their new odour to the traces of the old ones already clinging there.

The urine of tom-cats is notoriously strongly scented, so much so that even the inefficient human nostrils can detect it all too clearly. To the human nose it has a particularly unpleasant stench and many people are driven to having toms neutered in attempts to damp down this activity. Other cat odours are almost undetectable by humans. The glands on the head, for example, which are rubbed against objects to deposit another form of feline scent-mark, produce an odour that is of great significance to cats but goes completely unnoticed by the animals' human owners.

Some authorities have claimed that the squirted urine acts as a threat signal to rival cats. Hard evidence is lacking, however, and many hours of patient field-observation have never revealed any reactions to support this view. If the odour left on the landmarks was truly threatening to other cats, it should intimidate them when they sniff it. They should recoil in fear and panic and slink away. Their response is just the opposite. Instead of withdrawing, they are positively attracted to the scent-marks and sniff at them with great interest.

If they are not threatening, what do the territorial scent-marks signify? What signals do they carry? The answer is that they function rather as newspapers do for us. Each morning we read our papers and keep up to date with what is going on in the human world. Cats wander around their territories and, by sniffing at the scent-marks, can learn all the news about the comings and goings of the feline population. They can check how long it has been since their own last visit (by the degree of weakening of their own last scent-spray) and they can read the odour-signs of who else has passed by and sprayed, and how long ago. Each spray also carries with it considerable information about the emotional state and the individual identity of the sprayer. When a cat decides to have another spray itself, it is the feline equivalent of writing a letter to *The Times*, publishing a poem, and leaving a calling card, all rolled into one jet of urine.

It might be argued that the concept of scent-signalling is far-fetched and that urine-spraying by cats is simply their method of getting rid of waste products from the body and that it has no other significance whatsoever. If a cat has a full bladder it will spray; if it has an empty bladder it will not spray. The facts contradict this. Careful observation shows that cats perform regular spraying actions in a set routine, regardless of the state of their bladder. If it happens to be full, then each squirt is large. If it is nearly empty, then the urine is rationed out. The *number* of squirts and the territorial areas which are scent-marked remain the same, no matter how much or how little liquid the cat has drunk. Indeed, if the cat has completely run out of urine, it can be seen continuing its scent-marking routine, laboriously visiting each scent-post, turning its back on it, straining and quivering its tail, and then walking away. The act of spraying has its own separate motivation, which is a clear indication of its importance in feline social life.

Although it is not generally realized, females and neutered cats of both sexes do spray jets of urine, like tom-cats. The difference is that their actions are less frequent and their scent far less pungent, so that we barely notice it.

How large is a cat's territory?

The wild counterpart of the domestic cat has a huge territory, with males patrolling up to 175 acres. Domestic cats which have gone wild and are living in remote areas where there is unlimited space also cover impressively large areas. Typical farm cats use nearly as much space, the males ranging over 150 acres. Female farm cats are more modest, using only about fifteen acres on average. In cities, towns and suburbs, the cat population becomes almost as overcrowded as that of the human citizens. The territories of urban cats shrink to a mere fraction of the home range enjoyed by their country cousins. It has been estimated that cats living rough in London, for example, enjoy only about one-fifth of an acre each. Pampered pet cats living in their owners' houses may be even more restricted, depending on the size of the gardens attached to the houses. The maximum density recorded is one pet cat per one-fiftieth of an acre.

This degree of variation in the size of feline territories shows just how flexible the cat can be. Like people, it can adjust to a massive shrinkage of its home ground without undue suffering. From the above figures it is easy to calculate that 8,750 crowded pet cats could be fitted into the territory of one wild cat living in a remote part of the world. The fact that the social life of the crowded cats does not become chaotic and vicious is a testimony to the social tolerance of cats. In a way this is surprising, because people often speak of the sociability of dogs, but stress that cats are much more solitary and unsociable. They may be so by choice, but given the challenge of living whisker-by-tail with other cats, they manage remarkably well.

They achieve this high-density success in a number of ways. The most important factor is the provision of food by their owners. This removes the need for lengthy daily hunting trips. It may not remove the urge to set off on such trips – a well-fed cat remains a hunting cat – but it does reduce the determination born of an empty stomach. If they find themselves invading neighbouring territories, they can give up the hunt without starving. If restricting their hunting activities to their own cramped home ranges makes them inefficient prey-catchers, it might prove frustrating, but it does not lead to starvation and death. It has been demonstrated that the more food the cats are given by their owners, the smaller their urban territories become.

Another factor helping them is the way in which their human owners divide up their own territories – with fences and hedges and walls to demarcate their gardens. These provide natural boundary-lines that are easy to recognize and defend. In addition there is a permissible degree of overlapping in feline territories. Female cats often have special areas where several of their home ranges overlap and where they can meet on neutral ground. The males – whose territories are always about ten times the size of those of the females, regardless of how great or small the crowding – show much more overlap. Each male will roam about on an area that takes in several female territories, enabling him to keep a permanent check on which particular queen (female) is on heat at any particular moment.

The overlapping is permitted because the cats are usually able to avoid one another as they patrol the landmarks in their patch of land. If, by accident, two of them do happen to meet up unexpectedly, they may threaten one another or simply keep out of one another's way, watching each other's movements and waiting their turn to visit a particular zone of the territory.

The numbers of pet cats are, of course, controlled by their owners, with the neutering of adults, destruction of unwanted litters and the selling or giving away of surplus kittens. But how does the territorial arrangement of feral cats survive the inevitable production of offspring? One detailed study of dockland cats at a large port revealed that in an area of 210 acres there were ninety-five cats. Each year they produced about 400 kittens between them. This is a high figure of about ten per female, which must mean that on average each queen gave birth to two litters. In theory this would mean a fivefold increase in the population each year. In practice it was found that the population remained remarkably stable from one year to the next. The cats had established an appropriate territory size for the feral, dockland world in which they lived, and then kept to it. Closer investigation revealed that only one in eight of the kittens survived to become adults. These fifty additions to the population each year were balanced by fifty deaths among the older cats. The main cause of death here (as with most urban cat populations) was the fatal road accident.

How sociable are cats?

The cat is often characterized as a solitary, selfish animal, walking alone and coming together with other cats only to fight or mate. When cats are living wild, with plenty of space, it is true that they do fit this picture reasonably well, but they are capable of changing their ways when they become more crowded. Living in cities and towns, and in the homes of their human owners, cats show a remarkable and unexpected degree of sociability.

Anyone doubting this must remember that, to a pet cat, we ourselves are giant cats. The fact that domestic cats will share a home with a human family is, in itself, proof of their social flexibility. But this is only part of the story. There are many other ways in which cats demonstrate co-operation, mutual aid, and tolerance. This is particularly noticeable when a female is having kittens. Other females have been known to act as midwives, helping to chew through the umbilical cords and clean up the new-born offspring. Later they may offer a babysitting service, bring food for the new mother, and occasionally feed young from other litters as well as their own. Even males sometimes show a little paternal feeling, cleaning kittens and playing with them.

These are not usual activities, but despite the fact that they are uncommon occurrences, they do reveal that the cat is capable, under special circumstances, of behaving in a less selfish way than we might expect.

Territorial behaviour also involves some degree of restraint and sharing. Cats do their best to avoid one another, and often use the same ranges at different times as a way of reducing conflict. In addition there are special no-cat's-land areas where social 'clubs' can develop. These are parts of the environment where, for some reason, cats call a general truce and come together without too much fighting. This is common with feral city cats, where there may be a special feeding site. If humans throw food for them there, they may gather quite peacefully to share it. Close proximity is tolerated in a way that would be unthinkable in the 'home base' regions of these cats.

Considering these facts, some authorities have gone so far as to say that cats are truly gregarious and that their society is more co-operative than that of dogs, but this is romantic exaggeration. The truth is that, where social life is concerned, cats are opportunists. They can take it or leave it. Dogs, on the other hand, can never leave it. A solitary dog is a wretched creature. A solitary cat is, if anything, relieved to be left in peace.

If this is so, then how can we explain the mutual aid examples given above? Some are due to the fact that we have turned domestic cats into overgrown kittens. By continuing to feed them and care for them we prolong their juvenile qualities into their adult lives. Like Peter Pan, they never grow up mentally, even though they become mature adults physically. Kittens are playful and friendly with their litter-mates and with their mothers, so they are used to acting together in a small group. This quality can be added to later adult activities, making them less competitive and less solitary. Secondly, those cats living wild in cities, where there is little space, adapt to their shrunken territories out of necessity, rather than by preference.

Some animals can live only in close-knit social groups. Others can tolerate only a completely solitary existence. The cat's flexibility means that it can accept either mode of living, and it is this that has been a key factor in its long success story since it was first domesticated thousands of years ago.

Why do cats keep crying to be let out and then cry to be let in again?

Cats hate doors. Doors simply do not register in the evolutionary story of the cat family. They constantly block patrolling activities and prevent cats from exploring their home range and then returning to their central, secure base at will. Humans often do not understand that a cat needs to make only a brief survey of its territory before returning with all the necessary information about the activities of other cats in the vicinity. It likes to make these tours of inspection at frequent intervals, but does not want to stay outside for very long, unless there has been some special and unexpected change in the condition of the local feline population.

The result of this is an apparent perversity on the part of pet cats. When they are in they want to go out, and when they are out they want to come in. If their owner does not have a small cat-flap on the back door of the house, there will be a regular demand for attention, to assist the cat in its rhythmic territorial supervision. Part of the reason why this repeated checking of the outside world is so important is because of the time-clock message system of the scent-marks. Each time a cat rubs against a landmark in its territory or sprays urine on it, it leaves a personal scent which immediately starts to lose its power. This decline is at a steady rate and can be used by cats to determine how long it has been since the scent-marker rubbed or sprayed. The repeated visits by a cat to inspect its territory are motivated by a need to reactivate its fading scent signals. Once this has been done, comfort and security beckons again, and the anxious feline face appears for the umpteenth time at the window.

What does a cat signal with its ears?

Unlike humans, felines have very expressive ears. They not only change direction as the cat listens to sounds coming from different sources, but they also adopt special postures that reflect the emotional mood of the animals.

There are five basic ear signals, related to the following moods: relaxed, alert, agitated, defensive and aggressive.

In the relaxed cat the apertures of the ears point forward and slightly outward, as the animal quietly listens for interesting sounds over a wide range.

When the resting cat stirs itself and focuses on some exciting detail in its surroundings, the ear position changes into the 'alert mode'. As it stares at the point of interest, its ears become fully erect and rotate so that their apertures point directly forward. The ears are kept pricked in this way as long as the cat remains gazing straight ahead. The only variation occurs if there is a sudden noise away to the side of the animal, in which case an ear may be permitted a brief rotation in that direction without a shift of gaze.

An agitated cat, suffering from a state of conflict, frustration or apprehension, often displays a nervous twitching of the ears. In some species of wild cats this response has been made highly conspicuous by the evolution of long ear-tufts, but the domestic cat lacks this refinement and the ear-twitching itself is less common. Slight tufting does occur in some breeds, especially the Abyssinian where there is a small dark hairy point to the ear, but compared with the huge ear-tufts of a species such as the Caracal Lynx, this is a very modest development.

A defensive cat displays fully flattened ears. They are pressed tightly against the head as a way of protecting them during fights. The torn and tattered ears of battling tom-cats are a vivid testimony to the need to hide this delicate part of the anatomy as much as possible when the claws are out. The effect of flattening the ears to the sides of the head is to make them almost invisible when the animal is viewed from the front and to give its head a more rounded outline. There is one strange breed of cat called the Scottish Fold which has permanently flattened ears, giving it a continually defensive look, regardless of its true mood. What effect this has on its social life is hard to imagine.

An aggressive cat which is hostile without being particularly frightened has its own special ear posture. Here, the ears are rotated but not fully flattened. The *backs* of the ears become visible from the front, and this is the most dangerous ear signal any cat can transmit. In origin, this ear posture is half-way between alert and defensive – in other words, half-way between pricked forward and flattened backward. In effect, it is a 'ready for trouble' position. The animal is saying, 'I am ready to attack, but you don't frighten me enough to flatten my ears protectively.' The reason why this involves showing off the backs of the ears is because they must be rotated backwards before they can be fully flattened. So the rotated ears are in a 'ready-to-be-flattened' posture, should the aggressive cat's opponent dare to retaliate.

The aggressive ear posture has led to some attractive ear-markings in a number of wild cat species, especially the tiger, which has a huge white spot ringed with black on the back of each ear. When a tiger is angry, there is no doubt at all about its mood, as the pair of vivid white spots rotates into view. Again, domestic cats lack these special markings.

How do cats fight?

Under wild conditions cat-fights are a rarity because there is plenty of space, but in the more crowded urban and suburban areas feline territories become squashed together and frequently overlap. This means that a great deal of squabbling and serious physical duelling occurs, especially between rival tom-cats. Occasionally there are even killings or deaths resulting from battle injuries.

The primary objective of an attacking cat is to deliver a fatal neck-bite to its rival, employing much the same technique as when killing a prey. Because its opponent is of roughly the same size and strength, this lethal bite is hardly ever delivered. Indeed, the most craven and cowardly of rivals will defend itself to some extent, and a primary neck-bite is almost impossible to achieve.

The point to remember here is that even the most savage and dominant individual, as he goes in to the attack, is fearful of the consequences of the 'last-ditch-stand' by his terrorized underling. Driven into a corner, the weakling will try anything, lashing out with sharp claws and possibly injuring the dominant cat in a way that may pose a serious threat to his future hunting success and therefore his very survival. So even an out-and-out attacker shows fear mixed with his aggression, when the final crunch of physical contact arrives.

A typical sequence goes as follows: the dominant animal spots a rival and approaches it, adopting a highly characteristic threat posture, walking tall on fully stretched legs so that it suddenly appears bigger than usual. This effect is increased by the erecting of the hairs along its back. Because the crest is greater towards the rear end of the animal, the line of its back slopes up towards the tail. This gives the attacking cat a silhouette which is the exact opposite of the crouching shape of the weaker rival, whose rear end is held low on the ground.

With the backs of his ears showing and a great deal of howling, growling and gurgling, the attacker advances in slow motion, watching for any sudden reaction from his cringing enemy. The noises made are startlingly hostile and it is hard to understand how anything so totally aggressive can ever have been misnamed the tom-cat's 'love song'. One can only wonder at the love-life of the people who gave it this name. Needless to say, it has nothing whatever to do with true cat courtship.

As the attacking cat comes very near its rival, it performs a strange but highly characteristic head-twisting action. At a distance of about three feet it raises its head up slightly and then tilts it over to one side, all the time fixating the enemy with its eyes. Then the attacker takes a slow step forward and tilts its head the other way. This may be repeated several times and appears to be a threat of the neck-bite to come, the head twisting into the biting position as much as to say 'this is what you will get'. In other words, the attacker acts out the 'intention movement' of the assault typical of the species.

If two cats of equal status meet and threaten one another, a long period of deadlock may follow, with each animal performing exactly the same slow, hostile approach, as if displaying in front of a mirror. The nearer they get, the slower and shorter are their movements, until they become frozen in a prolonged stalemate which may last for many minutes. Throughout this they will continue to give vent to their caterwauling howls and moans, but neither side will be prepared to capitulate. Eventually they may separate from one another in incredibly slow motion. To increase their speed would be tantamount to admitting weakness and would lead to an immediate attack from the rival, so they must both withdraw with almost imperceptible movements to retain their status.

Should these threats and counter-threats collapse into a serious fight, the action begins with one of the adversaries making a lunging attempt at a neck-bite. When this happens the opponent instantly twists round and defends itself with its own jaws, while at the same time striking out with its front feet, clinging on with its forepaws and then kicking wildly with its powerful back feet. This is the point at which the 'fur flies' quite literally, and the growling gives way suddenly to yowls and screams as the two animals roll and writhe around, biting, clawing and kicking.

This phase does not last long. It is too intense. The rivals quickly pull apart and resume the threat displays, staring at one another and growling throatily once again. The assault is then repeated, perhaps several times, until one of them finally gives up and remains lying on the ground with its ears fully flattened. At this point the victor performs another highly characteristic display. It turns at right angles to the loser and, with great concentration, starts to sniff the ground, as though at that very moment there is an irresistibly delicious odour deposited there. The animal concentrates so hard on this sniffing that, were it not a regular feature of all fights, it would have the appearance of a genuine odour-check. But it is now only a ritual act, a victory display which signals to the cowering rival that its submission and capitulation have been accepted and that the battle is over. After the ceremonial sniffing the victor saunters slowly off and then, after a short while, the vanquished animal slinks away to safety.

Not all fights are conducted at such high intensity. Milder disputes are settled by 'paw-scrapping' in which the rivals swipe out at one another with extended claws. Slashing at their rival's head in this way, they may be able to settle their disagreement without the full ritual battle and all-in wrestling described above.

Why does a cat arch its back when it sees a strange dog?

If a cat feels threatened by a large dog, it pulls itself up on fully stretched legs and at the same time arches its back in the shape of an inverted U. The function of this display is clearly to make the cat look as big as possible, in an attempt to convince the dog that it is confronting a daunting opponent. To understand the origin of the display it is necessary to look at what happens when cats are threatening one another. If one cat is intensely hostile towards another and feels little fear, it approaches on stiffly stretched legs and with a straight back. If its rival is extremely frightened and feels no hostility, it arches its back and crouches low on the ground. In the case of the cat approached by a dog, there is both intense aggression *and* intense fear. It is this conflicting, double mood that gives rise to the special display. The cat borrows the most conspicuous element of its anger reaction – the stiff legs – and the most conspicuous element of its fear reaction – the arched back – and combines them to produce an 'enlarged cat' display. If it had borrowed the other elements – the straight back of anger and the low crouch of fear – the result would have been far from impressive.

Aiding its 'transformation display' is the fact that the animal, while stretching its legs and arching its back, also erects its fur and stands broadside-on to the dog. Together these four elements make up a compound display of maximum size increase. Even if the cat retreats a little, or advances towards the dog, it carefully keeps its broadside-on position, spreading its body in front of the dog like a bullfighter's cloak.

During the arched-back display the cat hisses ominously, like a snake, but this hissing turns to growling if it risks an attack. Then, when it actually lashes out at the dog, it adds an explosive 'spit' to its display. Experienced cats soon learn that the best policy when faced with a hostile dog is to go into the attack rather than run away, but it takes some nerve to do this when the dog is several times the cat's weight. The alternative of 'running for it' is much riskier, however, because once the cat is fleeing it triggers off the dog's hunting urges. To a dog a 'fleeing object' means only one thing – food – and it is hard to shift the canine hunting mood once it has been aroused. Even if the fleeing cat halts and makes a brave stand, it has little hope, because the dog's blood is up and it goes straight for the kill, arched back or no arched back. But if the cat makes a stand right from the first moment of the encounter with the dog, it has a good chance of defeating the larger animal, simply because by attacking it, the cat gives off none of the usual 'prey signals'. The dog, with sharp claws slashing at its sensitive nose, is much more likely then to beat a dignified retreat, and leave the hissing fury to its own devices. So, where dogs are concerned, the bolder the cat, the safer it is.

Why does a cat hiss?

It seems likely that the similarity between the hiss of a cat and that of a snake is not accidental. It has been claimed that the feline hiss is a case of protective mimicry. In other words, the cat imitates the snake to give an enemy the impression that it too is venomous and dangerous.

The quality of the hissing is certainly very similar. A threatened cat, faced with a dog or some other predator, produces a sound that is almost identical to that of an angry snake in a similar situation. Predators have great respect for venomous snakes, with good reason, and often pause long enough for the snake to escape. This hesitation is usually the result of an inborn reaction. The attacker does not have to learn to avoid snakes. Learning would not be much use in such a context, as the first lesson would also be the last. If a cornered cat is capable of causing alarm in an attacker by triggering off this instinctive fear of snakes, then it obviously has a great advantage, and this is probably the true explanation of the way in which the feline hiss has evolved.

Supporting this idea is the fact that cats often add spitting to hissing. Spitting is another way in which threatened snakes react. Also, the cornered cat may twitch or thrash its tail in a special way, reminiscent of the movements of a snake that is working itself up to strike or flee.

Finally, it has been pointed out that when a tabby cat (with markings similar to the wild type, or ancestral cat) lies sleeping, curled up tightly on a tree-stump or rock, its coloration and its rounded shape make it look uncannily like a coiled snake. As long ago as the nineteenth century it was suggested that the pattern of markings on a tabby cat are not direct, simple camouflage, but rather are imitations of the camouflage markings of a snake. A killer, such as an eagle, seeing a sleeping cat might, as a result of this resemblance, think twice before attacking.

Why does a cat wag its tail when it is hunting a bird on a lawn?

The scene is familiar to most cat-owners. Through the window they see their cat stalking a bird by creeping stealthily towards it, head down and body low on the ground. This cautious crouching attempt to be as inconspicuous as possible is suddenly and dramatically ruined by the animal's tail, which starts swishing uncontrollably back and forth through the air. Such movement acts like a flag being waved at the bird to warn it of approaching danger. The intended victim takes off immediately and flies to safety, leaving a frustrated feline hunter staring up into the sky.

Cat-owners witnessing this scene are puzzled by their cat's inefficiency. Why does the cat's tail betray the rest of the body in this self-defeating way? Surely the wild ancestors of the domestic cat could not have survived such a serious flaw in their hunting technique? We know that conspicuous tail-wagging in cats is a social signal indicating acute conflict. It is useful when employed between one cat and another and is then an important part of feline body language. But when it is transferred into a hunting context, where the only eyes that will spot the signal are those of the intended prey, it wrecks the whole enterprise. So why has it not been suppressed in such cases?

To find the answer we have to look at the normal hunting sequence of the cat. This does not take place on an open lawn and is less well known to cat-owners than it might be because it involves a great deal of waiting and hiding. If owners do happen on a hunt in progress they will automatically disrupt it, so that there is nothing more to observe. The disturbed prey escapes and the cat gives up. So for a casual observer the whole sequence is not easy to study. It requires some systematic and secretive catwatching. When this is undertaken, the following points emerge:

First, the cat makes a great deal of use of cover. It spends much time lying half-hidden in undergrowth, often with only its eyes and part of its face visible. The tail is usually completely hidden from view. Second, it never attempts to pounce on a prey until it is very close to it. It is not a prey-chaser. It may make a few stalking-runs, rushing forward in its flattened posture, but it then halts and waits again before pouncing. Third, its normal prey is not birds but rodents. A careful study of feral cats in the United States revealed that birds accounted for only 4 per cent of the diet. The excellent eyesight of birds and their ability to fly straight up in the air to escape make them unsuitable targets for domestic cats.

Together these points explain the dilemma of the suburban cat hunting a bird on a lawn. To start with, the open, manicured lawn robs the cat of all its natural cover, exposing its whole body to view. This is doubly damaging to its chances. It makes it almost impossible for the cat to creep near enough for its typical, close-quarters pounce without being seen. This puts it into an acute conflict between wanting to stay immobile and crouched, on the one hand, and wanting to rush forward and attack, on the other. The conflict starts its tail wagging furiously and the same lack of cover that created the conflict then cruelly exposes the vigorous tail movements to the frightened gaze of the intended prey.

If the attempt to hunt a bird on an open lawn is so doomed to failure, why does the cat keep trying? The answer is that every cat has a powerful urge to go hunting at regular intervals, but this urge has been severely hampered by advances in human pest control. In town and cities and suburbs, the rodent population that used to infest houses and other dwellings has been decimated by modern techniques. Garden birds may be pests, but their appeal to human eyes has protected them from a similar slaughter. As a result, the rodent-hunting cat finds itself today in an unnaturally mouse-free, bird-rich environment. It cannot utilize its natural hunting skills under such conditions. It is this state of affairs that drives the cat on to crouch hopelessly but compulsively on open lawns, staring longingly at elusive birds. So when it waves its tail at its prey on these occasions, it is not the cat which is a stupid hunter, but we who have unwittingly forced a clever hunter to attempt an almost impossible task.

Why does a cat chatter its teeth when it sees a bird through the window?

Not every owner has observed this curious action, but it is so strange that it is a case of 'once seen never forgotten'. The cat, sitting on a window-sill, spots a small bird conspicuously hopping about outside and stares at it intently. As it does so it begins juddering its teeth in a jaw movement which has variously been described as a 'tooth-rattling stutter', a 'tetanic reaction' and 'the frustrated chatter of the cat's jaws in the mechanical staccato fashion'. What does it mean?

This is what is known as a 'vacuum activity'. The cat is performing its highly specialized killing-bite, as if it already had the unfortunate bird clamped between its jaws. Careful observation of the way in which cats kill their prey has revealed that there is a peculiar jaw movement employed to bring about an almost instantaneous death. This is important to a feline predator because even the most timid of prey may lash out when actually seized, and it is vital for the cat to reduce as much as possible any risk of injury to itself from the sharp beak of a bird or the powerful teeth of a rodent. So there is no time to lose. After the initial pounce, in which the prey is pinioned by the strong claws of the killer's front feet, the cat quickly crunches down with its long canine teeth, aiming at the nape of the neck. With a rapid juddering movement of the jaws it inserts these canines into the neck, slipping them down between vertebrae to sever the spinal cord. This killing-bite immediately incapacitates the prey and it is an enactment of this special movement that the frustrated, window-gazing cat is performing, unable to control itself at the tantalizing view of the juicy little bird outside.

Incidentally, this killing-bite is guided by the indentation of the body outline of the prey – the indentation which occurs where the body joins the head in both small birds and small rodents. Some prey have developed a defensive tactic in which they hunch up their bodies to conceal this indentation and in this way make the cat miss its aim. If the trick works, the cat may bite its victim in part of the body which does not cause death, and on rare occasions the wounded prey may then be able to scrabble to safety if the cat relaxes for a moment, imagining that it has already dealt its lethal blow.

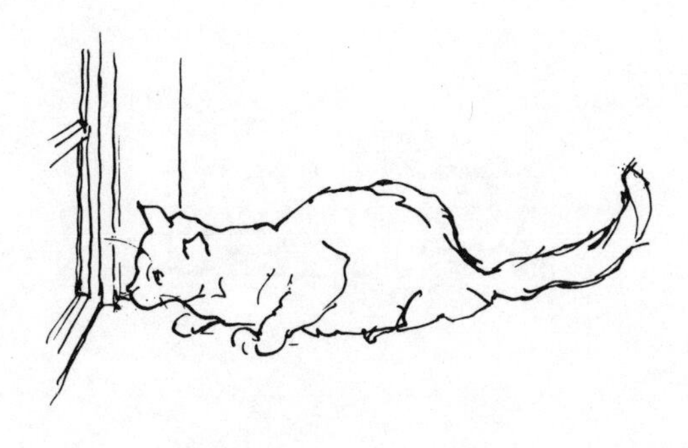

Why does a cat sway its head from side to side when staring at its prey?

When a cat is preparing to pounce on its prey it sometimes sways its head rhythmically from side to side. This is a device employed by many predators blessed with binocular vision. The head-sway is a method of checking the precise distance at which the prey is located. If you sway your own head from side to side you will see how, the closer an object is, the more it is displaced by the lateral movements. The cat does this to refine its judgment, because when the rapid pounce forward is made it must be inch-accurate or it will fail.

Why does a cat sometimes play with its prey before killing it?

Horrified cat-owners have often experienced the shock of finding their pet cats apparently torturing a mouse or small bird. The hunter, instead of delivering the killing-bite of which it is perfectly capable, indulges in a cruel game of either hit-and-chase or trap-and-release, as a result of which the petrified victim may actually die of shock before the final *coup de grâce* can be delivered. Why does the cat do it, when it is such an efficient killing machine?

To start with, this is not the behaviour of a wild cat. It is the act of a well-fed pet which has been starved of hunting activity as a consequence of the 'hygienic' environment in which it now lives – neat suburbs or well-kept villages where the rodent infestation has long since been dealt with by poison and human pest control agencies. For such a cat, the occasional catching of a little field mouse, or a small bird, is a great event. It cannot bear to end the chase, prolonging it as much as possible until the prey dies. The hunting drive is independent of the hunger drive – as any cat-owner knows whose cat has chased off after a bird on the lawn immediately after filling its belly with canned cat food. Just as hunger increases without food, so the urge to hunt increases without access to prey. The pet cat over-reacts, and the prey suffers a slow death as a result.

On this basis, one would not expect feral cats which are living rough or farm cats employed as 'professional pest-controllers' to indulge in play with their half-dead prey. In most cases it is indeed absent, but some researchers have found that female farm cats do occasionally indulge in it. There is a special explanation in their case. As females, they will have to bring live prey back to the nest to demonstrate killing to their kittens at a certain stage in the litter's development. This maternal teaching process will account for an eagerness on the part of females to play with prey even though they are not starved of the hunting process.

There is one other explanation for this seemingly cruel behaviour. When cats attack rats they are quite nervous of their prey's ability to defend themselves. A large rat can give a nasty bite to a cat and has to be subdued before any attempt is made to perform the killing-bite. This is done by the cat swinging a lightning blow with its claws extended. In quick succession it may beat a rat this way and that until it is dazed and dizzy. Only then does the cat risk going in close with its face for the killing-bite. Sometimes a hunting cat will treat a small mouse as if it were a threatening rat, and start beating it with its paws instead of biting it. In the case of a mouse this quickly leads to a disproportionately savage pounding, with the diminutive rodent being flung to and fro. Feline behaviour of this type may appear like playing with the prey, but it is distinct from the trap-and-release play and should not be confused with it. In trap-and-release play the cat is inhibiting its bite each time. It is genuinely holding back to prolong the hunt. In hit-and-chase attacks on mice the cat is simply over-reacting to the possible danger from the prey's teeth. It may look like cruel play, but in reality it is the behaviour of a cat that is not too sure of itself. Even after the prey is nearly or completely dead, such a cat may continue to bat the victim's body around, watching it intently to see if there is any sign of retaliation. Only after a long bout of this treatment will the cat decide that it is safe to deliver the killing-bite and eat the prey. An experienced, full-time hunter would not react in this way, but a pampered pet cat, being a little rusty on the techniques of a quick kill, may well prefer this safer option.

How does a cat prepare its food?

Immediately after the kill, a cat goes through the strange little routine of 'taking a walk'. Unless it is starving, it paces up and down for a while, as if feeling the need to release the tension of the hunt-and-kill sequence. Only then does it settle down to eating the prey. This pause may be important for the cat's digestion, giving its system a chance to calm down after the adrenalin-excitement of the moments that have just passed. During this pause a prey that has been feigning death may try to escape and, on very rare occasions, succeeds in doing so before the cat can return to the hunting mood again.

When the cat finally approaches its prey to eat it, there is the problem of how to prepare it for easy swallowing. Small rodents cause no difficulties. They are simply eaten head first and the skins, if swallowed, are regurgitated later. Some cats separate out the gall-bladder and intestines and avoid eating them, but others are too hungry to care and gobble down the entire animal without any fuss.

Birds are another matter because of their feathers, but even here the smaller species are eaten in their entirety, with the exception of tail and wing feathers. Birds the size of thrushes and blackbirds are plucked a little before eating, but then the cat impatiently starts its meal. After a while it breaks off to remove a few more feathers, before eating further. It repeats this a number of times as the feeding proceeds. Bigger birds, however, demand more systematic plucking, and if a cat is successful at killing a pigeon or something larger, it must strip away the feathers before it begins to eat.

To pluck a pigeon, a cat must first hold down the body of the bird with its front feet, seize a clump of feathers between its teeth, pull its jaw-clamped head upwards with some force, and then finally open its mouth and shake its head vigorously from side to side to remove any clinging plumage. As it shakes its head it spits hard and makes special licking-out movements with its tongue, trying to clear its mouth of stubbornly attached feathers. It may pause from time to time to lick its flank fur. This last action puts grooming into reverse. Normally the tongue cleans the fur, but here the fur cleans the tongue. Any last remnants are removed and then the next plucking action can take place.

The urge to pluck feathers from a large bird appears to be inborn. I once presented a dead pigeon to a wild cat living in a zoo cage where it had always been given chunks of meat as its regular diet. The cat became so excited at seeing a fully feathered bird that it started an ecstatic plucking session that went on and on until the whole body of the bird was completely naked. Instead of settling down to eat it, the cat then turned its attention to the grass on which it was sitting and began plucking that. Time and again it tugged out tufts of grass from the turf and shook them away with the characteristic bird-plucking movements until, eventually, having exhausted its long-frustrated urge to prepare its food, the cat finally bit into the flesh of the pigeon and began its meal. Clearly, plucking has its own motivation and can be frustrated by captivity, just like other, more obvious drives.

The strangest feature of feather-plucking is that Old World Cats perform it differently from New World Cats. All species from the first area perform a zigzag tugging movement leading to the full shake of the head, while those from the Americas tug the feathers out in a long vertical movement, straight up, and only then perform the sideways shake. It appears that, despite superficial similarities between the small cats from the two sides of the Atlantic, they are in reality two quite distinct groups.

How efficient is the cat as a pest-killer?

Before the cat became elevated to the level of a companion and pet for friendly humans, the contract between man and cat was based on the animal's ability to destroy pests. From the time mankind first started to keep grain in storage, the cat had a role to play and carried out its side of the bargain with great success.

Not so long ago it was thought that the best way to get farm cats to kill rats and other rodent pests was to keep the feline hunters as hungry as possible. This seemed obvious enough, but it was wrong. Hungry farm cats spread out over a huge hunting territory in search of food and killed fewer of the pests inside the farm. Cats that were fed by the farmer stayed nearer home and their tally of farm pests was much higher. The fact that they had been fed already and were not particularly hungry made no difference to the number of prey they killed each day, because the urge to hunt is independent of the urge to eat. Cats hunt for the sake of hunting. Once farmers realized this they were able to keep their cats close by the farm and reduce the damage done to their stores by rodent pests. A small group of farm cats, well looked after, could prevent any increase in the rodent population, providing a major infestation had not been allowed to develop before their arrival.

According to one authority, the champion mouser on record was a male tabby living in a Lancashire factory where, over a very long lifespan of twenty-three years, he killed more than 22,000 mice. This is nearly three a day, which seems a reasonable daily diet for a domestic cat, allowing for some supplements from human friends, but it is far exceeded by the world's champion ratter. That honour goes to a female tabby which earned her keep at the late lamented White City Stadium. Over a period of only six years she caught no fewer than 12,480 rats, which works out at a daily average of five to six. This is a formidable achievement and it is easy to see why the ancient Egyptians went to the trouble of domesticating cats and why the act of killing one was punishable by death.

Why do cats present freshly caught prey to their human owners?

They do this because they consider their owners such hopeless hunters. Although usually they look upon humans as pseudo-parents, on these occasions they view them as their family – in other words, their kittens. If kittens do not know how to catch and eat mice and small birds, then the cat must demonstrate to them. This is why the cats that most commonly bring home prey and offer such gifts to their owners are neutered females. They are unable to perform this action for their own litters, so they redirect it towards human companions.

The humans honoured in this way frequently recoil in horror or anger, especially if the small rodent or bird is still half-alive and struggling. The cat is totally nonplussed by this extraordinary response. If it is scolded for its generous act, it once again finds its human friends incomprehensible. The correct reaction would be to praise the cat for its maternal generosity, take the prey from it with many compliments and strokings and then quietly dispose of it.

Under natural conditions a cat which has a litter of kittens introduces them to prey animals little by little. When they are about seven weeks old, instead of killing and eating her prey where she catches it, she kills it and then brings it back to where the kittens are kept. There she proceeds to eat it while they watch. The next phase involves bringing the dead prey back and playing with it before consuming it, so that the kittens can see her beating it with her claws and grabbing it. The third phase involves leaving the prey to be eaten by the kittens themselves. But she is still not prepared to risk bringing a live or even a half-dead prey to the kittens, because it could easily bite them or attack them if they are unwary. Only when they are a little older will she do this, and then she herself will make the kill in front of the kittens. They watch and learn. Eventually they will accompany her on the hunt and try killing for themselves.

Why do cats eat grass?

Most cat-owners have observed the way in which, once in a while, their pet goes up to a long grass stem in the garden and starts to chew and bite at it. Cats living in apartments where there are no gardens in which to roam have been known to cause considerable damage to house plants in desperate attempts to find a substitute for grasses. In rare cases such cats have even harmed themselves by biting into plants that are poisonous.

Many cat experts have puzzled over this behaviour and some have admitted frankly that they have no answer. Others have offered a variety of explanations. For many years the favourite reply was that the cats use grass as a laxative to help them pass troublesome hairballs lodged in their intestines. A related suggestion claimed that they were eating grass to make themselves vomit up the hairballs. This was based on the observation that cats do sometimes vomit after eating grass, but it overlooked the possibility that whatever made the cats feel sick also made them want to eat grass, rather than that the grass-eating actually caused the vomiting.

A less popular explanation was that the grass aided the cats in the case of throat inflammation, or irritation of the stomach. Some authorities simply dismissed the activity as a way of adding roughage to the diet.

None of these explanations makes much sense. The amount of grass actually eaten is very small. Watching the cats chewing at long grasses, one gets the impression that they are merely taking in a little juice from the leaves and stems, rather than adding any appreciable solid bulk to their diet.

The most recent opinion – and the most likely explanation – is that cats chew grass to obtain minute quantities of a chemical substance that they cannot obtain from a meat diet and which is essential to their health. The substance in question is a vitamin called folic acid, and it is vital to cats because it plays an important role in the production of haemoglobin. If a cat is deficient in folic acid its growth will suffer and it may become seriously anaemic. Cat-owners whose animals have no access to grasses of any kind sometimes solve the problem by planting grass seeds in a tray and growing a patch of long grass in their apartments for their pets to chew on.

Incidentally, it is worth pointing out that although cats may need this plant supplement to their meat diet, they are first and foremost carnivores and must be treated as such. Recent attempts by well-meaning vegetarians to convert their cats to a meat-free diet are both misguided and cruel. Cats rapidly become seriously ill on a vegetarian diet and cannot survive it for long. The recent publication of vegetarian diets recommended as suitable for cats is a clear case of animal abuse and should be dealt with as such.

How does a cat use its whiskers?

The usual answer is that the whiskers are feelers that enable a cat to tell whether a gap is wide enough for it to squeeze through, but the truth is more complicated and more remarkable. In addition to their obvious role as feelers sensitive to touch, the whiskers also operate as air-current detectors. As the cat moves along in the dark it needs to manoeuvre past solid objects without touching them. Each solid object it approaches causes slight eddies in the air, minute disturbances in the currents of air movement, and the cat's whiskers are so amazingly sensitive that they can read these air changes and respond to the presence of solid obstacles even without touching them.

The whiskers are especially important – indeed vital – when the cat hunts at night. We know this from the following observations: a cat with perfect whiskers can kill cleanly both in the light and in the dark. A cat with damaged whiskers can kill cleanly only in the light; in the dark it misjudges its killing-bite and plunges its teeth into the wrong part of the prey's body. This means that in the dark, where accurate vision is impeded, healthy whiskers are capable of acting as a highly sensitive guidance system. They have an astonishing, split-second ability to check the body outline of the victim and direct the cat's bite to the nape of the unfortunate animal's neck. Somehow the tips of the whiskers must read off the details of the shape of the prey, like a blind man reading braille, and in an instant tell the cat how to react. Photographs of cats carrying mice in their jaws after catching them reveal that the whiskers are almost wrapped around the rodent's body, continuing to transmit information about the slightest movement, should the prey still be alive. Since the cat is by nature predominantly a nocturnal hunter, its whiskers are clearly crucial to its survival.

Anatomically the whiskers are greatly enlarged and stiffened hairs more than twice the thickness of ordinary hairs. They are embedded in the tissue of the cat's upper lip to a depth three times that of other hairs, and they are supplied with a mass of nerve-endings which transmit the information about any contact they make or any changes in air-pressure. On average the cat has twenty-four whiskers, twelve on each side of the nose, arranged in four horizontal rows. They are capable of moving both forwards, when the cat is inquisitive, threatening, or testing something, and backwards, when it is defensive or deliberately avoiding touching something. The top two rows can be moved independently of the bottom two, and the strongest whiskers are in rows two and three.

Technically whiskers are called vibrissae and the cat has a number of these reinforced hairs on other parts of its body – a few on the cheeks, over the eyes, on the chin and, surprisingly, at the backs of the front legs. All are sensitive detectors of movement, but it is the excessively long whiskers that are by far the most important vibrissae, and it is entirely apt that when we say that something is 'the cat's whiskers' we mean that it is rather special.

Why do cats' eyes glow in the dark?

Because they possess an image-intensifying device at the rear of their eyes. This is a light-reflecting layer called the *tapetum lucidum* (meaning literally 'bright carpet'), which acts rather like a mirror behind the retina, reflecting light back to the retinal cells. With this, the cat can utilize every scrap of light that enters its eyes. With our eyes we absorb far less of the light which enters them. Because of this difference cats can make out movements and objects in the semi-darkness which would be quite invisible to us.

Despite this efficient nocturnal ability it is not true that cats can see in complete darkness, as some people seem to believe. On a pitch black night they must navigate by sound, smell and the sensitivity of their amazing whiskers.

Why do cats' eyes contract to a vertical slit?

Reducing the pupils to slits, rather than tiny circles, gives the cat a more refined control over precisely how much light enters the eyes. For an animal with eyes sensitive enough to see in very dim light it is important not to be dazzled by bright sunlight, and the narrowing of the pupils to tight slits gives a greater and more accurate ability to cut down the light input. The reason why cats have vertical slits rather than horizontal ones is that they can use the closing of the lids to reduce the light input even further. With these two slits – the vertical one of the pupil and the horizontal one of the eyelids – working at right angles to one another, the feline eye has the possibility of making the most delicate adjustment of any animal, when faced with what would otherwise be a blinding light.

Confirmation of the fact that it is the nocturnal sensitivity of the cat's eyes that is linked with the contraction of the pupils to slits, is found in the observation that lions, which are daytime killers, have eyes that contract, like ours, to circular pinpricks.

Can cats see colours?

Yes, but rather poorly, is the answer. In the first half of this century scientists were convinced that cats were totally colour-blind and one authority reworked a popular saying with the words: 'Day and night, all cats see grey.' That was the prevailing attitude in the 1940s, but during the past few decades more careful research has been carried out and it is now known that cats can distinguish between certain colours, but not, apparently, with much finesse.

The reason why earlier experiments failed to reveal the existence of feline colour vision was because in discrimination tests cats quickly latched on to subtle differences in the degree of greyness of colours and then refused to abandon these clues when they were presented with two colours of exactly the same degree of greyness. So the tests gave negative results. Using more sophisticated methods, recent studies have been able to prove that cats can distinguish between red and green, red and blue, red and grey, green and blue, green and grey, blue and grey, yellow and blue, and yellow and grey. Whether they can distinguish between other pairs of colours is still in dispute. For example, one authority believes that they can also tell the difference between red and yellow, but another does not.

Whatever the final results of these investigations one thing is certain: colour is not as important in the lives of cats as it is in our own lives. Their eyes are much more attuned to seeing in dim light, where they need only one-sixth of the light we do to make out the same details of movement and shape.

How does a female cat deal with her new-born kittens?

As the nine-week gestation period comes to an end the pregnant cat becomes restless, searching around for a suitable den or nest in which to deliver her kittens. She looks for somewhere quiet, private and dry. In a house, strange noises emanate from cupboards and other nooks and crannies as the cat tests out a variety of suitable sites. Suddenly, from being increasingly ravenous, her hunger vanishes and she refuses food, which means that the moment of birth is imminent – perhaps only a few hours away. At this point she disappears and settles down to the serious business of bringing a litter of kittens into the world.

Some cats hate interference at this stage and become upset by too much attention. Others – usually those that have never been given much privacy in the house – do not seem to care much one way or the other. The happy-go-lucky ones will co-operatively move into a specially prepared birth-box, with soft warm bedding provided and easy accessibility for a human midwife, should one be needed. Other cats stubbornly refuse the perfect nest-bed offered them and perversely disappear into the shoe-cupboard or some such dark, private place.

Giving birth is a lengthy process for the average cat. With a typical litter of, say, five kittens, and with a typical delay of, say, thirty minutes between the arrival of each one, the whole process lasts for two hours, after which both cat and kittens are quite exhausted. Some cats give birth much more quickly – one kitten per minute – but this is rare. Others may take as long as an hour between kittens – but this is also uncommon. The typical time delay of about half an hour is not an accident. It gives the mother long enough to attend to one kitten before the next arrives.

The attention she gives the new-born baby consists of three main phases. First, she breaks away the birth sac (the amniotic sac) which encases the kitten as it emerges into the world. She then pays special care to the cleaning of the nose and mouth of the new-born, enabling it to take its first breath. Once this crucial stage is over, she starts to clean up, biting through the umbilical cord and eating it, up to about one inch from the kitten's belly. The little stump she leaves alone, and this eventually dries out and finally drops off of its own accord. She then eats the afterbirth – the placenta – which provides her with valuable nourishment to see her through the long hours of total kitten-caring that now face her, during their first day of life. After this she licks the kitten all over, helping to dry its fur, and then she rests. Soon the next kitten will appear and the whole process will have to be repeated. If she grows tired, towards the end of an unusually large litter, the last one or two kittens may be ignored and left to die, but most female cats are amazingly good midwives and need no help from their human owners.

As the kittens recover from the trauma of birth, they start rooting around, searching for a nipple. The first feed they enjoy is vitally important because it helps to immunize them against disease. Before she produces her full-bodied nutritional milk, the mother provides a thin first-milk called *colostrum*, which is rich in antibodies and gives the kittens an immediate advantage in the coming struggle to avoid the diseases of infancy. It is also rich in proteins and minerals and its production lasts for several days, before the mother cat starts to produce the normal milk supply.

How do kittens avoid squabbling when feeding from the mother?

Within a few days of birth each kitten has developed an attachment to its own personal nipple, which it recognizes with ease. Amazing as it may sound, this is possible because each nipple has a special smell. We know this because if the belly region of the mother cat is washed by her human owner, so that it is cleansed of its natural fragrance, the kittens fail to find their favourite nipples. Instead of peacefully taking up their usual stations, they become disorientated. Confusion reigns and squabbling occurs.

It is remarkable to think that in the 'simple' world of the very young kitten there is odour detection based on differences so subtle that they can label each nipple as clearly as name-cards on school lockers, and that in this way orderly sharing can be maintained at feeding time.

At what rate do kittens develop?

When they are born the kittens are blind and deaf, but have a strong sense of smell. They are also sensitive to touch and soon start rooting for the mother's nipples. At this stage they weigh between two and four ounces, the average birth weight being roughly three-and-a-half ounces. They are about five inches long.

By day four, the kittens have already started the paw-treading action which helps to stimulate the mother's milk-flow. At the end of the first week of life their eyes begin to open and they have by now doubled their body weight. As they approach the end of their first month of life, they show the first signs of playing with one another. They can move themselves about with more efficiency and can sit up. Whatever colour their eyes will be later in life, at this stage all kittens are blue-eyed and remain so until they are about three months old. Their teeth are beginning to break through at the age of one month.

At roughly thirty-two days, they eat their first solid food, but they will not be weaned until they are two months old. (Wild cats take longer to wean their kittens – about four months.) During their second month of life they become very lively and intensely playful with one another. Inside the house, pet kittens will use their mother's dirt tray by the time they are one-and-a-half months old. Play-fighting and play-hunting become dominant features at the end of the second month.

In their third month of life they are in for a shock. The mother refuses to allow them access to her nipples. They must now make do entirely with solids and with liquids lapped from a dish. Before long their mother will be coming into oestrus again and concentrating on tom-cats once more.

In their fifth month the young cats begin to scent-mark their home range. They are shedding their milk teeth and exploring their exciting new world in a less playful manner. The chances are that their mother is already pregnant again by now, unless her human owners have kept her indoors against her will.

At six months, the young cats are fully independent, capable of hunting prey and fending for themselves.

Why does a cat move its kittens to a new nest?

When the kittens are between twenty and thirty days old, their mother usually moves them to a new nest site. Each kitten is picked up firmly by the scruff of the neck and, with mother's head held as high as possible, is carried off to the fresh location. If it has to be transported over a long distance the mother may grow tired of the weight and let her head sag, switching from carrying to dragging. The kitten never objects, lying limp and still in its mother's jaws, with its tail curled up between its bent hind legs. This posture makes the kitten's body as short as possible and reduces the danger of bumps as it is unceremoniously shunted from old nest to new one.

As soon as the mother arrives at the new site she has chosen, she opens her jaws and the kitten drops to the ground. She then returns for the next kitten and the next, until the whole litter has been transported. After the last one has been moved, she makes a final trip to inspect the old nest, making doubly sure that nobody has been left behind. This suggests that counting kittens is not one of the cat's strong points.

It is usually stated that this removal operation is caused either by the old nest becoming fouled or because the kittens have outgrown it. These explanations seem logical enough, but they are not the true reason. A cat with a large, clean nest is just as likely to set about moving its litter. The real answer lies with the wild ancestors of the domestic cat. In the natural environment, away from canned cat food and dishes of milk, the mother cat must start bringing prey back to the nest, to arouse the carnivorous responses of her offspring. When the kittens are between thirty and forty days old they will have to begin eating solids, and it is this change in their behaviour that is behind the removal operation. The first, old nest had to be chosen for maximum snugness and security. The kittens were so helpless then and needed protection above all else. But during the second month of their lives, after their teeth have broken through, they need to learn how to bite and chew the prey animals brought by the mother. So a second nest is needed to facilitate this. The primary consideration now is proximity to the best food supply, reducing the mother's task of repeatedly bringing food to her young.

This removal operation still occurs in domestic cats – if they are given half the chance – despite the fact that the feeding problem has been eliminated by the regular refilling of food dishes by their human owners. It is an ancient pattern of maternal feline behaviour which, like hunting itself, refuses to die away simply because of the soft lifestyle of domestication.

In addition to this 'food-source removal pattern', there are, of course, many examples of a cat quickly transporting her litter away from what she considers to be a dangerous nest site. If human curiosity becomes too strong and prying eyes and groping hands cannot keep away from the 'secret' nest, strange human smells may make it an unattractive abode. The mother cat may then search for a new home, simply to get more privacy. Moves of this kind can take place at any stage of the maternal cycle. In wild species of cats, interference with the young at the nest may result in a more drastic measure, the mother refusing to recognize them as offspring any more, and abandoning them or even eating them. What happens, in effect, is that the alien smells on the kitten's body make it into an alien 'species' – in other words, into a prey species – and the obvious response to such an object is to eat it. Domestic cats rarely respond in this way, because they have become so used to the scents and odours of their human owners that they do not class them as alien. Kittens handled by humans therefore usually remain 'in the family', even if they have acquired new scents.

How do kittens learn to kill?

The short answer is that they do not need to learn how to perform the killing action, but it does help if they get some instruction from their mother. Kittens reared by scientists, in isolation from the mother cat, were able to kill prey when given live rodents for the first time. Not all these kittens succeeded, however. Out of twenty tested, only nine killed and only three of those actually ate their kills. Kittens reared in a rodent-killing environment, where they could witness kills but never saw the prey eaten, were much more successful. Eighteen out of twenty-one such kittens tested were killers and nine of these actually ate their kills.

Interestingly, of eighteen kittens reared in the company of rodents, only three became rodent-killers later on. The other fifteen could not be trained to kill later by seeing other cats killing. For them the rodents had become 'family' and were no longer 'prey'. Even the three killers would not attack rodents of the same species as the one with which they were reared. Although it is clear that there is an inborn killing pattern with kittens, this pattern can be damaged by unnatural rearing conditions.

Conversely, really efficient killers have to experience a kittenhood which exposes them to as much hunting and killing as possible. The very best hunters are those which, as youngsters, were able to accompany their mother on the prowl and watch her dealing with prey. Also, at a more tender age, they benefited from her bringing prey to the nest to show them. If the mother does not bring prey to the kittens in the nest between the sixth and twentieth week of their lives, they will be far less efficient as hunters in later life.

Why does a kitten sometimes throw a toy into the air when playing?

The scene is familiar enough. A kitten tires of stalking and chasing a ball. It suddenly and without warning flips one of its paws under the ball, flinging it up into the air and backwards over its head. As the ball flies through the air, the kitten swings round and follows it, pouncing on it and 'killing' it yet again. As a slight variation, faced with a larger ball, it will perform the backward flip using both front feet at the same time.

The usual interpretation of this playful behaviour is that the kitten is being inventive and cunningly intelligent. Because its toy will not fly up into the air like a living bird, the kitten 'puts life into it' by flinging the ball over its shoulder, so that it can then enjoy pursuing the more excitingly 'lively' prey-substitute. This credits the kitten with a remarkable capacity for creative play – for inventing a bird in flight. In support of this idea is the fact that no adult cat hunting a real bird would use the 'flip-up' action of the front paws. This action, it is argued, is the truly inventive movement, reflecting the kitten's advanced intelligence.

Unfortunately this interpretation is wrong. It is based on an ignorance of the instinctive hunting actions of the cat. In the wild state, cats have three different patterns of attack, depending on whether they are hunting mice, birds or fish. With mice, they stalk, pounce, trap with the front feet and then bite. With birds they stalk, pounce and then, if the bird flies up into the air, they leap up after it, swiping at it with both front feet at once. If they are quick enough and trap the bird's body in the pincer movement of their front legs, they pull it *down* to the ground for the killing-bite. Less familiar is the way in which cats hunt for fish. They do this by lying in wait at the water's edge and then, when an unwary fish swims near, they dip a paw swiftly into the water and slide it rapidly under the fish's body, flipping the fish up out of the water. The direction of the flip is back and over the cat's shoulders, and it flings the fish clear of the water. As the startled fish lands on the grass behind the cat, the hunter swings round and pounces. If the fish is too large to be flipped with the claws of just one front foot, then the cat may risk plunging both front feet into the water at once, grabbing the fish from underneath with its extended claws and then flinging the prey bodily backwards over its head.

It is these instinctive fishing actions that the kittens are performing with their 'flip-up' of the toy ball, not some new action they have learned or invented. The reason why this has been overlooked in the past is because few people have watched cats fishing successfully in the wild, whereas many people have seen their pets leaping up at birds on the garden lawn.

A Dutch research project was able to reveal that the scooping up of fish from the water, using the 'flip-up' action, matures surprisingly early and without the benefit of maternal instruction. Kittens allowed to hunt fish regularly from their fifth week of life onwards, but in the absence of their mother, became successful anglers by the age of seven weeks. So the playful kitten throwing a ball over its shoulder is really doing no more than it would do for real, if it were growing up in the wild, near a pond or river.

When do cats become sexually mature?

This occurs when they are nearly a year old, but there is a great deal of variation. For toms, the youngest recorded age for sexual maturity is six months, but this is abnormal. Eight months is also rather precocious, and the typical male does not become sexually active until he is between eleven and twelve months of age. For toms living rough, it may be considerably longer – more like fifteen to eighteen months, probably because they are given less chance in the competition with older males.

For females, the period can be relatively short, six to eight months being usual, but very young females only three to five months old have been known to come into sexual condition. This early start seems to be caused by the unnatural circumstances of domestication. For a wild cat, ten months is more usual.

The European Wild Cat, for instance, starts its breeding season in March. There is a gestation period of sixty-three days and then the kittens appear in May. By late autumn they strike off on their own and, if they survive the winter, they will themselves start to breed the following March when they are about ten months old, producing their own litters when they are a year old. For these wild cats there is only one season a year, so young toms may have to be patient and wait for the following season before they go into action.

This wild cycle is obviously geared to the changing seasons and the varying food supply, but for the pampered pet there are no such problems. With its hunting ears finely tuned to the metallic sound of a can-opener and with the central-heating humming gently in the background, the luxuriating house cat has little to fear from the annual cycle of nature. As a result, its breeding sequences are less rigid than its wild counterpart. It may breed as early as the second half of January, producing a litter by the beginning of April. Two months later, with its kittens weaned and despatched to new homes, it may well start off again with another breeding sequence, producing a second litter in the late summer. With this loss of a simple annual rhythm, there is a whole scatter of ages among young domestic cats, leading to the variations in the stages at which they become sexually active.

Cases have been reported of wild cats producing a second litter in August, but it is suspected that this only occurs where there has been interbreeding between the wild animals and feral domestic cats.

How fast do cats breed?

Without restraint a cat population can increase at a startling rate. This is because female cats are excellent mothers, and because domestication has led to a possible tripling of the number of litters and to an increase in litter size. European Wild Cats, with their single litter each year, have an average of two to four kittens, but domestic cats may produce an average of four to five kittens in each of their three annual litters.

A simple calculation, starting with a single breeding pair of domestic cats, and allowing for a total of fourteen kittens in each three-litter year, reveals that in five years' time there will be a total of 65,536 cats. This assumes that all survive, that males and females are born in equal numbers and that they all start breeding when they are a year old. In reality, the females might start a little younger, so the figure could be higher. But against this is the obvious fact that many would perish from disease or accident.

This paints a grim picture for the aspiring house mouse, a nightmare world of wall-to-wall cats. But it never materializes because there are enough responsible human owners to ensure that breeding restraints *are* applied to their pet cats, to keep the numbers under control. Neutering of both males and females is now commonplace and it is estimated that more than 90 per cent of all toms have suffered the operation. Females that are allowed to breed may have their litter size reduced to one or two, the unfortunate kittens being painlessly killed by the local vet. In some areas there are fairly ruthless extermination programmes for feral and stray cats, and in certain countries there have even been oral contraceptive projects, with the stray cat population given food laced with 'the pill'. Israel, for example, claims to prevent about 20,000 kittens a year by using this technique.

Despite these attempts, however, there are still well over a million feral and stray cats in Great Britain at the present time. It has been estimated that there are as many as half a million in the London region alone. In addition there are between four and five million pet cats, making a massive feline population of roughly one cat per ten humans.

How do cats perform their courtship?

Cats spend a great deal of time building up to the mating act, and their prolonged 'orgies' and promiscuity have given them a reputation for lasciviousness and lust, over the centuries. This is not because the mating act itself is lengthy or particularly erotic in form. In fact, the whole process of copulation rarely exceeds ten seconds and is often briefer than that. What gives the felines their reputation for lechery is the superficial resemblance between their sexual gatherings and a Hell's Angels gang-bang. There is a female, spitting and cursing and swiping out at the males one moment and writhing around on the ground the next. And there is a whole circle of males, all growling and howling and snarling at one another as they take it in turns (apparently) to rape the female.

The truth is slightly different. Admittedly, the process may take hours, even days, of non-stop sexual activity, but it is the female who is very much in charge of what is happening. It is she who calls the tune, not the males.

It begins when the female comes on heat and starts calling to the males. They also respond to her special sexual odours and are attracted from all around. The male on whose territory she has chosen to make her displays is initially strongly favoured, because other males from neighbouring territories will be scared at invading his ground. But a female on heat is more than they can resist, so they take the risk. This leads to a great deal of male-to-male squabbling (and accounts for most of the noise – which is why the caterwauling and howling is thought of, mistakenly, as sexual, when in reality it is purely aggressive). But the focus of interest is the female and this helps to damp down the male-to-male fighting and permits the gathering of a whole circle of males around her.

She displays to them with purring and crooning and rolling on the ground, rubbing herself and writhing in a manner that fascinates the male eyes fixating upon her. Eventually one of the males, probably the territory owner himself, will approach her and sit close to her. For his pains, he is attacked with blows from her sharp-clawed forepaws. She spits and growls at him and he retreats. Any male approaching her too soon is seen off in this way. She is the mistress of the situation and it is she who will eventually choose which male may approach her more intimately. The male who succeeds in this may or may not be the dominant tom present. That is up to her. But certain male strategies do help the toms to succeed. The most important one is to advance towards her only when she is looking the other way. As soon as she turns in his direction the male freezes – like a child playing the party game called 'statues'. She only attacks when she sees the actual advance itself, not the immobile body that has somehow, by magic, come a little nearer. In this way a tom with finesse can get quite close. He offers her a strange little chirping noise and, if she gives up spitting and hissing at him, he will eventually risk a contact approach. He starts by grabbing the scruff of her neck in his jaws and then carefully mounts her. If she is ready to copulate she flattens the front of her body and raises her rump up into the air, twisting her tail to one side. This is the posture called 'lordosis' and is the final invitation signal to the male, permitting him to copulate.

As time goes on, the 'orgy' changes its style. The males become satiated and are less and less interested in the female. She, on the other hand, seems to become more and more lustful. Having worked her way through one male after another at comparatively short intervals for perhaps several days, one might imagine that she too would be satiated, but this is not so. As long as her peak period of heat persists she will want to be mated, and the toms now have to be encouraged. Instead of playing hard to get, she now has to work on the males to arouse their interest. She does this with a great deal of crooning, rubbing and especially writhing on the ground. The males still sit around watching her, and from time to time manage to muster enough enthusiasm to mount her once more. Eventually it is all over and the chances of a female cat returning home unfertilized after such an event are utterly remote.

Why does the tom grab the female by the scruff of the neck when mating?

At first sight this appears to be a piece of macho brutality, in similar vein to the cartoon caveman grabbing his mate by her hair and dragging her off to his cave. Nothing could be further from the truth. In sexual matters it is the female, not the male, who is dominant, where cats are concerned. Toms may fight savagely among themselves, but when they are sexually aroused and attempting to mate with the queen they are far from bossy. It is the female who swipes out and beats the toms. The bite on the back of her neck may look savage, but in reality it is a desperate ploy on the part of the male to protect himself from further assault. This protection is of a special kind. It is not a matter of forcibly holding down the female so that she cannot twist round and attack him. She is too strong for that. Instead it is a 'behaviour trick' played by the male. All cats, whether male or female, retain a peculiar response to being grabbed firmly by the scruff of the neck, dating back to their kitten days. Kittens have an automatic reaction to being held in this way by their mother. She uses it when it is necessary to transport the kittens from an unsafe to a safe place. It is crucially important that the kittens do not struggle on such occasions, where their very lives may be at stake. So felines have evolved a 'freeze' reaction to being taken by the scruff of the neck – a response which demands that they stay quite still and do not struggle. This helps the mother in her difficult task of moving the litter to safety. When they grow up, cats never quite lose this response, as you can prove to yourself by holding an adult pet cat firmly by the skin of its neck. It immediately stops moving and will remain immobile in your grasp for some time before becoming restless. If you grasp it firmly on some other part of its body the restlessness is much quicker to occur, if not instantaneous. This 'immobilization reaction' is the trick the toms apply to their potentially savage females. The females are so claw-happy that the toms badly need such a device. As long as they hang on with their teeth, they have a good chance that the females will be helplessly transformed into 'kittens lying still in their mother's jaws'. Without such a behaviour trick the tom would return home with even more scars than usual.

Why does the female scream during the mating act?

As the tom finishes the brief act of copulation, which lasts only a few seconds, his female twists round and attacks him, swiping out savagely with her claws and screaming abuse at him. As he withdraws his penis and dismounts he has to move swiftly, or she is liable to injure him. The reason for her savage reaction to him at this point is easy to understand if you examine photographs of his penis taken under the microscope. Unlike the smooth penis of so many other mammals, the cat's organ is covered in short, sharp spines, all pointing away from the tip. This means that the penis can be inserted easily enough, but when it is withdrawn it brutally rakes the walls of the female's vagina. This causes her a spasm of intense pain and it is this to which she reacts with such screaming anger. The attacked male, of course, has no choice in the matter. He cannot adjust the spines, even if he wishes to do so. They are fixed and, what is more, the more sexually virile the male, the bigger the spines. So the sexiest male causes the female the most pain.

This may sound like a bizarre sado-masochistic development in feline sex, but there is a special biological reason for it. Human females who fail to become pregnant ovulate at regular intervals, regardless of whether they have mated with a male or not. Human virgins, for example, ovulate month after month, but this is not the case with cats. A virgin cat would not ovulate at all. Cats *only* ovulate after they have been mated by a male. It takes a little while – about twenty-five to thirty hours – but this does not matter because the intense period of heat lasts at least three days, so she is still actively copulating when ovulation occurs. The trigger that sets off the ovulation is the intense pain and shock the queen feels when her first suitor withdraws his spiny penis. This violent moment acts like the firing of a starting pistol which sets her reproductive hormonal system in operation.

In a way, it is not far from the truth to call a female cat on heat 'masochistic' because, within about thirty minutes of having been hurt by the first male penis, she is actively interested in sex again and ready to be mated once more, with a repeat performance of the scream-and-swipe reaction. Considering how much the spiny penis must have hurt her, it is clear that in a sexual context there is one kind of pain which does not produce the usual negative response.

Why do cats sneer?

Every so often a cat can be seen to pause and then adopt a curious sneering expression, as if disgusted with something. When first observed, this reaction was in fact called an 'expression of disgust' and described as the cat 'turning up its nose' at an unpleasant smell, such as urine deposited by a rival cat.

This interpretation is now known to be an error. The truth is almost the complete opposite. When the cat makes this strange grimace, known technically as the *flehmen* response, it is in reality appreciating to the full a delicious fragrance. We know this because tests have proved that urine from female cats in strong sexual condition produces powerful grimacing in male cats, while urine from females not in sexual condition produces a much weaker reaction.

The response involves the following elements: the cat stops in its tracks, raises its head slightly, draws back its upper lip and opens its mouth a little. Inside the half-opened mouth it is sometimes possible to see the tongue flickering or licking the roof of the mouth. The cat sniffs and gives the impression of an almost trancelike concentration for a few moments. During this time it slows its breathing rate and may even hold its breath for several seconds, after sucking in air. All the time it stares in front of it as if in a kind of reverie.

If this behaviour were to be likened to a hungry man inhaling the enticing smells emanating from a busy kitchen, it would not be too far from the truth, but there is an important difference. For the cat is employing a sense organ that we sadly lack. The cat's sixth sense is to be found in a small structure situated in the roof of the mouth. It is a little tube opening into the mouth just behind the upper front teeth. Known as the vomero-nasal or Jacobsen's organ, it is about half an inch long and is highly sensitive to airborne chemicals. It can best be described as a taste-smell organ and is extremely important to cats when they are reading the odour-news deposited around their territories. During human evolution, when we became increasingly dominated by visual input to the brain, we lost the use of our Jacobsen's organs, of which only a tiny trace now remains, but for cats it is of great significance and explains the strange, snooty, gaping expression they adopt occasionally as they go about the social round.

How does a cat manage to fall on its feet?

Although cats are excellent climbers they do occasionally fall, and when this happens a special 'righting reflex' goes into instant operation. Without this a cat could easily break its back.

As it starts to fall, with its body upside-down, an automatic twisting reaction begins at the head end of the body. The head rotates first, until it is upright, then the front legs are brought up close to the face, ready to protect it from impact. (A blow to a cat's chin from underneath can be particularly serious.) Next, the upper part of the spine is twisted, bringing the front half of the body round in line with the head. Finally, the hind legs are bent up, so that all four limbs are now ready for touchdown and, as this happens, the cat twists the rear half of its body round to catch up with the front. Finally, as it is about to make contact, it stretches all four legs out towards the ground and arches its back, as a way of reducing the force of the impact.

While this body-twisting is taking place, the stiffened tail is rotating like a propeller, acting as a counterbalancing device. All this occurs in a fraction of a second and it requires slow-motion film to analyse these rapid stages of the righting response.

How do cats behave when they become elderly?

Many owners fail to notice that their cats have reached 'old age'. This is due to the fact that senility has little effect on the feline appetite. Because they continue to eat greedily and with their usual vigour, it is thought that they are still 'young cats'. But there are certain tell-tale signs of ageing. Leaping and grooming are the first actions to suffer, and for the same reason. Old age makes the cat's joints stiffer and this leads to slower movements. Leaping up on to a chair or a table, or outside up on to a wall, becomes increasingly difficult. Very old cats actually need to be lifted up on to a favourite chair. As the supple quality of the young cat's flexible body is lost, it also becomes increasingly awkward for the cat to twist its neck round to groom the more inaccessible parts of its coat. These areas of its fur start to look dishevelled and at this stage a little gentle grooming by the animal's owner is a great help, even if the cat in question is not one that has generally been fussed over with brush and comb in its younger days.

As the elderly cat's body becomes more rigid, so do its habits. Its daily routine becomes increasingly fixed and novelties cause distress now, where once they may have aroused acute interest. The idea of buying a young kitten to cheer up an old cat simply does not work. It upsets the elderly animal's daily rhythm. Moving house is even more traumatic. The kindest way to treat an elderly cat is therefore to keep as much as possible to the well-established pattern of the day, but with a little physical help where required.

The outdoor life of an elderly cat is fraught with dangers. It has reached a point where disputes with younger rivals are nearly always going to end in defeat for the old animal, so a close eye must be kept on any possible persecution.

Luckily these changes do not occur until late in the lives of most cats. Human beings suffer from 'old age' for roughly the last third of their lives, but with cats it is usually only the last tenth. So their declining years are mercifully brief. The average lifespan is reckoned to be about ten years. Some authorities put it a little higher, at about twelve years, but it is impossible to be accurate because conditions of cat-keeping vary so much. The best rough guide is to say that a domestic cat should live between nine and fifteen years and should only suffer from senile decline for about the final year or so of that span.

There have been many arguments about the record longevity of a domestic cat, with some amazing claims being made, some as high as forty-three years. The longest accepted lifespan at present, however, is thirty-six years for a tabby called 'Puss' which lived from 1903 to 1939. This is exceptional and extremely rare. Serious attempts to locate cats over twenty years old, both in Great Britain and the United States, have never managed to turn up more than a mere handful of reliable cases.

One of the reasons why it is difficult to find good records of long-lived cats is that the most carefully kept details are always for the pure-bred animals, which live much shorter lives than the crossbred 'moggies' or mongrels. This is because the prized and carefully recorded pure-bred cats suffer from inbreeding which shortens their lives. The 'badly bred' alley cat, by comparison, enjoys what is called 'hybrid vigour' – the improved physical stamina produced by outbreeding. Unfortunately, such cats are less well looked after in most cases, so they, in their turn, suffer more from fighting, neglect and irregular diet. This cuts down their lifespan. The cat with a record-breaking lifespan is therefore most likely to be the one which has a dubious pedigree but is a much loved and protected pet. For such an animal fifteen to twenty years is not too hopeless a target.

One of the strangest features of cat longevity is that it easily exceeds that of dogs. The record for a dog is twenty-nine years, seven years short of the longest-lived cat. Bearing in mind the fact that larger animals usually live longer than smaller ones, the figures should be reversed, so for their size cats do unusually well. And there is a compensation for those toms that suffer the mutilation of being neutered, for neutered toms have a longer lifespan than 'entire' ones. The reasons for this, it appears, are that they get involved in fewer damaging brawls with rivals, and also that they are, for some reason, more resistant to infection. One careful study revealed that a neutered tom could expect on average three more years of feline survival than an unneutered one.

Why does a cat lick its face when it is not dirty?

A quick flick of the tongue over the lips is one of the tell-tale signs that a cat is becoming agitated, while at the same time being fascinated or puzzled by something. Keeping an eye firmly fixed on the source of its agitation, the cat gives the impression that it has suddenly and inexplicably developed an urgent need to clean its nose or the fur around its mouth. But there is no dirt there. The cleaning is not functional and it does not follow the usual pattern seen after feeding or during a normal grooming session. The licks are short and sharp – rapid sweeps of the tongue that do not develop in the usual way into proper washing actions. They are the cat's equivalent of a man scratching his head when perplexed or irritated.

Reactions of this type are called 'displacement activities'. They occur when the cat is thrown into a state of conflict. Something which upsets it but at the same time arouses its curiosity, will simultaneously repel and attract the animal. There it sits, wanting to leave and wanting to stay. It stares at the irritant and, unable to resolve its conflict, shows its state of agitation by performing some trivial, abbreviated action – anything to break the stalemate in which it finds itself. Different species respond in different ways. Some animals nibble their paws, others scratch behind an ear with a hind leg. Birds wipe their beaks on a branch. Chimpanzees scratch their arms or their chins. But for felines the tongue-swipe is the favourite action.

There is a harmless way in which this can be tested. Cats do not like vibrating noises with a high pitch, but they are intrigued by what makes such sounds. A coin rubbed back and forth along the teeth of an ordinary hair-comb produces such a noise. Almost every cat, when hearing the brrrrrr sound produced by this action, stares at the comb in your hand and then, after a few seconds, starts licking its lips. If the sound continues, the animal may eventually decide that it has had enough and it will get up and walk away. Amazingly, this works for fully grown lions just as well as for small tabbies. Sometimes the lip-licking gives way to a violent sneeze, sometimes to a wide yawn. These actions appear to be alternative feline 'displacement activities', but they are less common than the lip-licking.

Why a cat is so irritated by a vibration sound is something of a mystery unless, during the course of feline evolution, it has come to represent a noxious animal of some kind – something unsuitable to attack as a prey. An obvious example that comes to mind here is the rattlesnake's rattle. Cats perhaps have an automatic alarm response to such animals and this may account for the fact that they are upset and yet intrigued at the same time.

Why do cats react so strongly to catnip?

In a word, it is because they are junkies. The catnip plant, a member of the mint family, contains an oil called *hepetalactone*, an unsaturated lactone which does for some cats what marijuana does for some people. When cats find this plant in a garden they take off on a ten-minute 'trip' during which they appear to enter a state of ecstasy. This is a somewhat anthropomorphic interpretation because we have no idea what is really happening inside the cat's brain, but anyone who has seen a strong catnip reaction will know just how trancelike and drugged the animal seems to become. All species of cats react in this way, even lions, but not every individual cat does so. There are some non-trippers and the difference is known to be genetic. With cats, you are either born a junkie or you are not. Conditioning has nothing to do with it. Under-age cats, incidentally, never trip. For the first two months of life all kittens avoid catnip, and the positive reaction to it does not appear until they are three months old. Then they split into two groups – those that no longer actively avoid catnip, but simply ignore it and treat it like any other plant in the garden, and those that go wild as soon as they contact it. The split is roughly 50/50, with slightly more in the positive group.

The positive reaction takes the following form: the cat approaches the catnip plant and sniffs it; then, with growing frenzy, it starts to lick it, bite it, chew it, rub against it repeatedly with its cheek and its chin, head-shake, rub it with its body, purr loudly, growl, miaow, roll over and even leap in the air. Washing and clawing are also sometimes observed. Even the most reserved of cats seems to be totally disinhibited by the catnip chemical.

Because the rolling behaviour seen during the trancelike state is similar to the body actions of female cats in oestrus, it has been suggested that catnip is a kind of feline aphrodisiac. This is not particularly convincing, because the 50 per cent of cats that show the full reaction include both males and females, and both entire animals and those which have been castrated or spayed. So it does not seem to be a 'sex trip', but rather a drug trip which produces similar states of ecstasy to those experienced during the peak of sexual activity.

Cat junkies are lucky. Unlike so many human drugs, catnip does no lasting damage, and after the ten-minute experience is over the cat is back to normal with no ill-effects.

Catnip (*Nepeta cataria*) is not the only plant to produce these strange reactions in cats. Valerian (*Valeriana officinalis*) is another one, and there are several more that have strong cat-appeal. The strangest discovery, which seems to make no sense at all, is that if catnip or valerian are administered to cats internally they act as tranquillizers. How they can be 'uppers' externally and 'downers' internally remains a mystery.

How does a cat find its way home?

Over short distances each cat has an excellent visual memory, aided when close to home by familiar scents. But how does a cat manage to set off in the right direction when it is deliberately taken several miles away from its home territory?

First of all, can it really do this? Some years ago a German zoologist borrowed a number of cats from their owners, who lived in the city of Kiel. He placed them in covered boxes and drove them round and round the city, taking a complex, winding route to confuse them as much as possible. Then he drove several miles outside town to a field in which he had installed a large maze. The maze had a covered central area with twenty-four passages leading from it. Looked at from above, the passages fanned out like compass points, at intervals of fifteen degrees. The whole maze was enclosed, so that no sunlight or starlight could penetrate to give navigation clues to the cats. Then each cat in turn was placed in the maze and allowed to roam around until it chose an exit passage. In a significant number of cases, the cats selected the passage which was pointing directly towards their home.

When these findings were first reported at an international conference most of us present were highly sceptical. The tests had certainly been rigorously conducted, but the results gave the cats a homing sensitivity so amazing that we found it hard to accept. We suspected that there must be a flaw in the experimental method. The most obvious weakness was the possibility of a memory map. Perhaps the cats could make allowances and corrections for all the twists and turns the van took as it drove around the city, so that throughout the journey they kept recalculating the direction of their home base.

This doubt was removed by some other tests with cats done in the United States. There, the cats were given doped food before the trip, so that they fell into a deep, drugged sleep throughout the journey. When they arrived they were allowed to wake up fully and were then tested.

Astonishingly, they still knew the way home.

Since then many other navigation tests have been carried out with a variety of animals and it is now beyond doubt that many species, including human beings, possess an extraordinary sensitivity to the earth's magnetic field which enables them (and us) to find the way home without visual clues. The experimental technique which clinched this was one in which powerful magnets were attached to the navigators. This disrupted their homing ability.

We are still learning exactly how this homing mechanism works. It seems likely that iron particles occurring naturally in animal tissues are the vital clue, giving the homing individuals a built-in biological compass. But there is clearly a great deal more to discover.

At least we can now accept some of the incredible homing stories that have been told in the past. Previously they were considered to be wildly exaggerated anecdotes, or cases of mistaken identity, but now it seems they must be treated seriously. Cases of cats travelling several hundred miles to return from a new home to an old one, taking several weeks to make the journey, are no longer to be scoffed at.

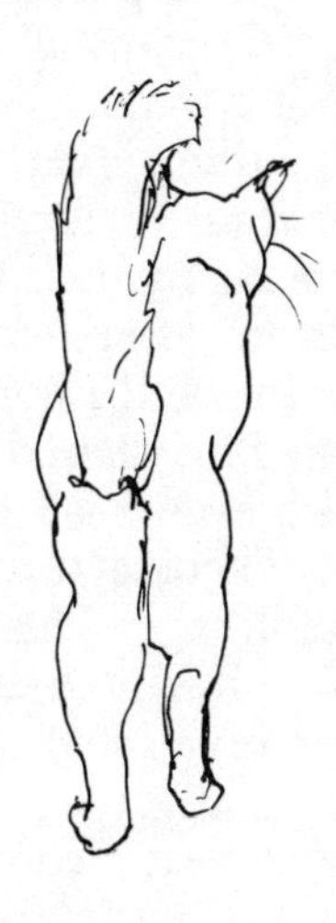

Can cats predict earthquakes?

The short answer is yes, they can, but we are still not sure how they do it. They may be sensitive to vibrations of the earth so minute that our instruments fail to detect them. It is known that there is a gradual build-up to earthquakes, rather than one sudden, massive tremor. It may be that cats have an advance warning system.

A second possibility is that they are responsive to the dramatic increase in static electricity that apparently precedes earthquakes. In humans there is also a response to these changes, but it is rather vague and unspecific. We speak of tenseness or throbbing in the head on such occasions, but we cannot distinguish these feelings from times when we have had a stressful day at work or perhaps when we are coming down with a cold. So we cannot read the signs accurately. In all probability cats can.

A third explanation sees cats as incredibly responsive to sudden shifts in the earth's magnetic field. Shifts of this type accompany earthquakes. Perhaps all three reactions occur at once – detection of minute tremors, electrostatic activity and magnetic upheavals. One thing is certain, cats have repeatedly become intensely agitated just before major earthquakes have struck. Cat-owners recognizing their pets' fears may well owe them their lives. In many cases cats have been observed suddenly rushing about inside the house, desperate to escape. Once the doors are opened for them they flee in panic from the buildings. Some females even rush back and forth carrying their kittens to safety. Then, a few hours later, the quake strikes and levels the buildings. This has been reported time and again from the most vulnerable earthquake areas and now serious research is under way to analyse precisely which signals the cats receive.

Similar responses have been recorded when cats have predicted volcanic eruptions or severe electrical storms. Because of their exceptional sensitivity they have often been foolishly credited with supernatural powers. In medieval times this was frequently their undoing, and many cats met a horrible death by burning at the hands of superstitious Christians because they appeared to be possessed of 'unnatural knowledge'. The fact that we now know this knowledge to be wholly natural makes it no less marvellous.

Why do we talk about catnaps?

Because cats indulge in brief periods of light sleep so frequently. In fact, these short naps are so common in cats and so rare in healthy humans that it is not exaggerating to say that cats and people have a fundamentally different sleep pattern. Unless human adults have been kept awake half the night, or are sick or extremely elderly, they do not indulge in brief naps. They limit their sleeping time to a single prolonged period of approximately eight hours each night. By comparison, cats are super-sleepers, clocking up a total, in twenty-four hours, of about sixteen hours, or twice the human period. This means that a nine-year-old cat approaching the end of its life has only been awake for a total of about three years. This is not the case in most other mammals and puts the cat into a special category – that of the refined killer. The cat is so efficient at obtaining its highly nutritious food that it has evolved time to spare, using this time to sleep and, apparently, to dream. Other carnivores, such as dogs and mongooses, have to spend much more time scurrying round, searching and chasing. The cat sits and waits, stalks a little, kills and eats, and then dozes off like a well-fed gourmet. Nothing falls asleep quite so easily as a cat.

There are three types of feline sleep: the brief nap, the longer light sleep and the deep sleep. The light sleep and the deep sleep alternate in characteristic bouts. When the animal settles down for more than a nap, it floats off into a phase of light sleep which lasts for about half an hour. Then the cat sinks further into slumber and, for six to seven minutes, experiences deep sleep. After this it returns to another bout of thirty minutes of light sleep, and so on until it eventually wakes up. During the periods of deep sleep the cat's body relaxes so much that it usually rolls over on to its side and this is the time when it appears to be dreaming, with frequent twitchings and quivering of ears, paws and tail. The mouth may make sucking movements and there are even occasional vocalizations, such as growls, purrs and general mutterings. There are also bursts of rapid eye movement, but throughout all this the cat's trunk remains immobile and totally relaxed. At the start of its life, as a very young kitten during its first month, it experiences only this deep type of sleep which lasts for a total of about twelve hours out of every twenty-four. After the first month the kittens rapidly switch to the adult pattern.

Why are cat-owners healthier than other people?

This may sound a strange question, but there is a great deal of evidence to prove that cat-owning is good for your health. And there is some comfort for beleaguered pet-owners, often criticized today for 'messing up the environment with their animals', that the anti-pet lobby will die younger than they will.

There are two reasons for this. First, it is known that the friendly physical contact with cats actively reduces stress in their human companions. The relationship between human and cat is touching in both senses of the word. The cat rubs against its owner's body and the owner strokes and fondles the cat's fur. If such owners are wired up in the laboratory to test their physiological responses, it is found that their body systems become markedly calmer when they start stroking their cats. Their tension eases and their bodies relax. This form of feline therapy has been proved in practice in a number of acute cases where mental patients have improved amazingly after being allowed the company of pet cats.

We all feel somehow released by the simple, honest relationship with the cat. This is the second reason for the cat's beneficial impact on humans. It is not merely a matter of touch, important as that may be. It is also a matter of psychological relationship which lacks the complexities, betrayals and contradictions of human relationships. We are all hurt by certain human relationships from time to time, some of us acutely, others more trivially. Those with severe mental scars may find it hard to trust again. For them, a bond with a cat can provide rewards so great that it may even give them back their faith in human relations, destroy their cynicism and their suspicion and heal their hidden scars. And a special study in the United States has recently revealed that, for those whose stress has led to heart trouble, the owning of a cat may literally make the difference between life and death, reducing blood pressure and calming the overworked heart.

Why is a female cat called a queen?

Because when she is on heat she lords it over the toms. They must gather around her like a circle of courtiers, must approach her with great deference, and are often punished by her in an autocratic manner.

Why is a male cat called a tom?

This can be traced back precisely to the year 1760 when an anonymous story was published called *The Life and Adventures of a Cat*. In it the 'ram cat', as a male was then known, was given the name 'Tom the Cat'. The story enjoyed great popularity, and before long anyone referring to a male cat, instead of calling it a 'ram', used the word Tom, which has survived now for over 200 years.

Why is a brothel called a cathouse?

Prostitutes have been called cats since the fifteenth century, for the simple reason that the urban female cat attracts many toms when she is on heat and mates with them one after the other. As early as 1401, men were warned of the risks of chasing the 'cattis tailis', or cat's tail. This also explains why the word 'tail' is sometimes used today as slang for female genitals. A similar use for the word 'pussy' dates from the seventeenth century.

Why do we say 'he let the cat out of the bag'?

The origin of this phrase, meaning 'he gave away a secret', dates back to the eighteenth century when it referred to a market-day trick. Piglets were often taken to market in a small sack, or bag, to be sold. The trickster would put a cat in a bag and pretend that it was a pig. If the buyer insisted on seeing it, he would be told that it was too lively to risk opening up the bag, as the animal might escape. If the cat struggled so much that the trickster let the cat out of the bag, his secret was exposed. A popular name for the bag itself was a 'poke', hence that other expression 'never buy a pig in a poke'.

Why do we speak of not having a 'cat-in-hell's' chance?

At first sight this is as puzzling as the well-known footballer's lament of being as 'sick as a parrot'. In both cases the mystery is solved if you know the original, unabbreviated saying which has long since been discontinued. The complete cat phrase is: 'No more chance than a cat in hell without claws'. It was originally a reference to the hopelessness of being without adequate weapons. (The original parrot saying, incidentally, was 'as sick as a parrot with a rubber beak' – a similar allusion to the lack of a sharp weapon.)

Why do we speak of someone 'having kittens'?

When we say 'she will have kittens if she finds out about this' we mean that someone will be terribly upset, almost to the point of hysteria. At first sight there is no obvious connection between distraught human behaviour and giving birth to kittens. True, a panic-stricken or hysterical woman who happens to be pregnant might suffer a miscarriage as a result of the intense emotional distress, so suddenly giving birth as a result of panic is not hard to understand. But why kittens? Why not puppies, or some other animal image?

To find the answer we have to turn the clock back to medieval times, when cats were thought of as the witch's familiars. If a pregnant woman was suffering agonizing pains, it was believed that she was bewitched and that she had kittens clawing at her inside her womb. Because witches had control over cats, they could provide magical potions to destroy the litter, so that the wretched woman would not give birth to kittens. As late as the seventeenth century an excuse for obtaining an abortion was given in court as removing 'cats in the belly'.

Since any woman believing herself to be bewitched and about to give birth to a litter of kittens would become hysterical with fear and disgust, it is easy to see how the phrase 'having kittens' has come to stand for a state of angry panic.

Why does a cat have nine lives?

The cat's resilience and toughness led to the idea that it had more than one life, but the reason for endowing it with nine lives, rather than any other number, has often puzzled people. The answer is simple enough. In ancient times nine was considered a particularly lucky number because it was a 'trinity of trinities' and therefore ideally suited for the 'lucky' cat.

Why do we say 'there is no room to swing a cat'?

This refers to the whip called 'the cat', employed on early naval vessels, and not to the animal itself. The cat, or cat-o'-nine-tails (because it has nine separate knotted thongs), was too long to swing below decks. As a result, sailors condemned to be punished with a whipping had to be taken up above, where there *was* room to swing a cat.

The reason why the whip itself was called a cat was because it left scars on the backs of the whipped sailors reminiscent of the claw marks of a savage cat.

Why do we say 'it is raining cats and dogs'?

This phrase became popular several centuries ago at a time when the streets of towns and cities were narrow, filthy and had poor drainage. Unusually heavy storms produced torrential flooding which drowned large numbers of the half-starved cats and dogs that foraged there. After a downpour was over, people would emerge from their houses to find the corpses of these unfortunate animals, and the more gullible among them believed that the bodies must have fallen from the sky – and that it had literally been raining cats and dogs.

A description of the impact of a severe city storm, written by Jonathan Swift in 1710, supports this view: 'Now from all parts the swelling kennels flow, and bear their trophies with them as they go ... drowned puppies, stinking sprats, all drenched in mud, dead cats, and turnip-tops, come tumbling down the flood.'

Some classicists prefer a more ancient explanation, suggesting that the phrase is derived from the Greek word for a waterfall: *catadupa*. If rain fell in torrents – like a waterfall – then the saying 'raining catadupa' could gradually have been converted into 'raining cats and dogs'.